3-31-64
6-9-64

THE DYNAMIC METHOD IN OCEANOGRAPHY

Elsevier Oceanography Series

THE DYNAMIC METHOD
IN OCEANOGRAPHY

BY

L. M. FOMIN

*Institute of Oceanology, Academy of
Sciences of the U.S.S.R., Moscow*

ELSEVIER PUBLISHING COMPANY
AMSTERDAM-LONDON-NEW YORK
1964

TRANSLATED BY
SCRIPTA TECHNICA INC., WASHINGTON, D.C., U.S.A.

EDITED BY
THOMAS WINTERFELD, NATIONAL OCEANOGRAPHIC DATA CENTER

ELSEVIER PUBLISHING COMPANY
335 JAN VAN GALENSTRAAT, P.O. BOX 211, AMSTERDAM

AMERICAN ELSEVIER PUBLISHING COMPANY, INC.
52 VANDERBILT AVENUE, NEW YORK 17, N.Y.

ELSEVIER PUBLISHING COMPANY LIMITED
12B RIPPLESIDE COMMERCIAL ESTATE
RIPPLE ROAD, BARKING, ESSEX

LIBRARY OF CONGRESS CATALOG CARD NUMBER 64-14174
WITH 57 ILLUSTRATIONS AND 23 TABLES

ALL RIGHTS RESERVED
THIS BOOK OR ANY PART THEREOF MAY NOT BE REPRODUCED IN ANY FORM,
INCLUDING PHOTOSTATIC OR MICROFILM FORM,
WITHOUT WRITTEN PERMISSION FROM THE PUBLISHERS

PRINTED IN THE NETHERLANDS

Contents

Introduction ix

Chapter I. Theoretical Principles of Indirect Computation of Sea Currents 1
1. Dynamic Method of Computing Sea Currents. Critical Comments on the Views of N. N. Zubov and H. Sverdrup 1
2. Wind-Driven Currents in an Infinite Channel. Pure Drift and Gradient Components of the Velocity of a Wind-Driven Current. Role of Friction in a Gradient Current. Mechanism of the Mutual Adjustment of the Density and Current Velocity Fields in the Sea 13
3. Some Considerations Concerning Convective Components of Current Velocity in the Sea. Currents Induced by a Difference in Climatic Conditions over the Ocean and by the Transport of Heat and Salt in a Wind-Driven Current 24
4. Baroclinic Layer of the Sea. Dependence of the Depth of the Lower Boundary of a Current on Water Stratification and on the Wind Field over the Sea 30

Chapter II. Accuracy of the Dynamic Method of Computing Ocean Currents 47
5. Statement of the Problem 47
6. Accuracy of Sea Water Temperature Measurements 48
7. Accuracy of Sea Water Salinity (Chlorinity) Determinations 49
8. Accuracy of Sea Water Density Computations . 52

9. Accuracy of Dynamic Height Computations... 55
10. Accuracy of Current Velocity Computation by the Dynamic Method... 63
11. Discussion of Results... 68
    Conclusions... 71

Chapter III. Investigation of the Vertical Velocity Distribution of a Wind-Driven Current in a Deep Sea Using a Density Model.. 73
12. Statement of the Problem... 73
13. Method of Computing Deep Sea Currents from the Surface Current and Atmospheric Pressure Gradient... 75
14. Investigation of the Depth of Penetration of a Wind-Driven Ocean Current... 101
15. Method of Increasing the Accuracy of Dynamic Computation... 113
    Conclusions... 116

Chapter IV. Methods for Computing the "Zero" Surface in the Sea... 117
16. General Evaluation of Methods for Determining the Layer of no Motion in the Sea... 117
17. Identification of the Layer of no Motion in the Sea with the Intermediate Oxygen Minimum (Dietrich's Method)... 118
18. Determination of the Layer of no Motion in the Sea from the Distortion of the Thickness of Isopycnal Layers (Parr's Method)... 124
19. Determination of the Depth of the Layer of no Motion from Salinity Distribution (Hidaka's Method)... 129
20. Computation of the Vertical Distribution of Current Velocity by the Dynamic Method Using Continuity Equations (Hidaka's Method)... 133
21. Determination of the Depth of the "Zero" Surface from the Analysis of Differences in the Dynamic Depths of Isobaric Surfaces (Defant's Method)... 139
    Conclusions... 148

Chapter V. Practical Procedures for Computing Currents by the Dynamic Method 149
22. Computation of Currents in Shallow Areas by the Dynamic Method 149
23. Computation of Currents by the Dynamic Method Relative to a Nonhorizontal "Zero" Surface 155
24. Absolute Computation of Gradient Current Velocity by the Dynamic Method Under Conditions Satisfying the Law of the Parallelism of Solenoids in the Sea 169
Conclusions 172

Chapter VI. Dynamic Chart of the Sea Surface in the Kurile-Kamchatka Region of the Pacific Ocean 174
25. Determination of the "Zero" Surface 174
26. Construction of a Dynamic Sea Surface Chart 195
Conclusions 201

Resume 202

Bibliography 204

Index 210

Introduction

A method of computing ocean currents from the distribution of water density was proposed in 1903 by I. Sandström and B. Helland-Hansen. This method became known later as the dynamic method of hydrological data analysis or the dynamic method for computing ocean-current elements.

The method is an indirect one, since it does not consider explicitly the forces that induce currents and are their primary cause. It is based on the assumption that in the steady state the horizontal pressure gradient and the coriolis force are in equilibrium in the sea below the mixed surface layer. Current velocities are computed by the dynamic method from the water density distribution which adjusts itself to the current system while developing and becoming established.

Strictly speaking, only relative horizontal pressure gradients and the difference between current velocities at two horizons can be computed from the distribution of water density in the sea. The horizon, or surface, in the sea relative to which current velocities are computed by the dynamic method and from which dynamic heights and depths are read is usually called the reference surface. The current velocities computed by the dynamic method describe the motion of water relative to water particles at the reference surface.

To obtain the absolute velocity of the gradient current by the dynamic method one must either know the distribution of gradient current velocity at the reference surface or place this surface in a layer where the horizontal pressure gradient and the geostrophic component of velocity are equal to zero or are insignificantly small. If the last condition is satisfied (i.e., there is no gradient current at the reference surface), this surface is usually called the "zero" surface. It is quite evident that the computation of current velocities by the dynamic method depends on the correct selection of the "zero" surface in an ocean or a sea.

As compared to other methods of studying water circulation in the sea, the dynamic method has the advantage of not involving complex and cumbersome computations. Also, the basic data are much easier to obtain than direct measurements of current velocities at sea. Numerous examples of applying the dynamic method show that the computations give, as a rule, a correct idea of water circulation which agrees with other data and with direct measurements. But there are cases when the dynamic method yields negative results.

Even though computation of currents by the dynamic method is simple, construction of a dynamic chart or of a hydrological profile involves original research requiring an individual intelligent consideration of the actual conditions in each case. It is very important that the investigator keep in mind the various possibilities of the dynamic method and the assumptions on which it is based.

An extensive literature is devoted to problems concerning the application of the dynamic method to the computation of current components. Some papers discuss the results of current computations using the dynamic method while others examine methodological problems in shallow areas, the determination of the "zero" surface, or computation of currents relative to a reference surface of complex form. The dynamic method of analyzing measurements at sea is most adequately described in the pamphlet of N. N. Zubov and O. I. Mamaev (1956) which discusses its principles and contains a number of valuable suggestions and recommendations on its practical application.

But some papers contain unfortunate inaccuracies and even faulty interpretations of the applicability of the dynamic method. This interferes with a correct understanding of theoretical principles and reduces its practical effectiveness.

This book is an attempt to justify the dynamic method theoretically and to elucidate the nature of currents computed by it. The accuracy of dynamic computations and the critical importance of the depth of the reference surface are examined. The structure of the gradient component of the velocity of a wind-driven current, the principal component computed by the dynamic method, is analyzed and a method is proposed for determining the depth below which the water density field does not noticeably affect the velocity of the gradient current. A special chapter is devoted to a critical review of methods for determining the "zero" surface. The main practical procedures for computing currents by the dynamic method under complex conditions, such as in shallow areas and with a nonhorizontal "zero" surface, are examined. All the major results are illustrated in the last chapter, where a dynamic chart of the surface of the northwestern Pacific is constructed.

The author has not attempted to present a thorough treatise on the dynamic method, but has merely tried to analyze some aspects of its theoretical principles and practical application, using the theory of ocean currents.

The author wishes to express his sincere gratitude to the members of the Laboratory of Ocean Dynamics of the Institute of Oceanography, Academy of Sciences of the USSR, where this book was written and published.

CHAPTER I

Theoretical Principles of Indirect Computation of Sea Currents

1. DYNAMIC METHOD OF COMPUTING SEA CURRENTS. CRITICAL COMMENTS ON THE VIEWS OF N. N. ZUBOV AND H. SVERDRUP

On the basis of Bjerknes' circulation theorem (1900), Sandström and Helland-Hansen (1903) obtained a formula making it possible to compute current velocities at particular horizons from the distribution of water density without direct measurement.

$$v_1 - v_2 = \frac{D_a - D_b}{2\omega L \sin \varphi}. \tag{1.1}$$

In this expression $(v_1 - v_2)$ is the difference between average current velocities at two horizons; ω is the angular speed of the earth's rotation; L is the distance between two hydrological stations a and b at which the vertical distribution of sea water density is known from observation; φ is the average geographic latitude of the area; and D is the dynamic height of the isobaric surface p_1 relative to the isobaric surface p_2 given by

$$D = \int_{p_2}^{p_1} \alpha \, dp, \tag{1.2}$$

where α is the specific volume of sea water and p is pressure.

Computation of current velocity by these formulas became known as the dynamic method of analyzing data, or the dynamic method of computing the components of sea currents. Without dwelling on the well-known derivation of formulas for the dynamic method, let us note that this method does not consider the influence of frictional forces on fluid motion in the steady state. The coriolis force that arises during motion is considered in this method as being in equilibrium with the horizontal pressure gradient. In other words, the current computed by the dynamic method is assumed to be geostrophic.

First, we must have a clear idea of what components of the velocity of an ocean current are being computed by the dynamic method from the distribution of water density. Otherwise, the pattern of vertical current velocity variations and the effect of individual factors on this pattern would be useless. The general derivation of dynamic method formulas does not indicate what component of current velocity is being computed by this method. In spite of the fact that the dynamic method has existed and has been widely used for half a century in oceanographic practice, views are varied and contradictory.

N. N. Zubov (1929) derived a formula similar to formula (1.1) from simple geometrical considerations. The simplified approach to the theory of the method promoted to a large extent its practical application (at the present time the dynamic method is the standard method of analysis). This apparent simplicity creates the impression that the dynamic method is universal. V. B. Shtokman was the first to point this out (1937).

The fact is, that by stating the general problem of the relationship between the field of sea water density and the current velocity field, N. N. Zubov based his arguments on a particular example that best fits the gradient current in a channel. This creates the impression that the density field, because of its adjustment to the current system, is an indicator of the velocity field, regardless of what forces induce water circulation in the sea. This was the reason why the dynamic method was frequently used mechanically without regard to specific circumstances. N. N. Zubov maintains that the velocity of a current of any origin can be computed from the density field provided

1. DYNAMIC METHOD OF COMPUTING SEA CURRENTS

that such a current "occupies a stable position conforming to the configuration of the bottom, the shores, and neighboring current systems" (Zubov and Mamaev, 1956).

There is another opinion. In discussing the pressure field in the sea, H. Sverdrup (1942) represents the horizontal pressure gradient as the sum of two components, one caused by the slope of the sea surface and not changing in the vertical direction and the other created by the disturbance of the sea water density field and changing along the vertical. Sverdrup believes, accordingly, that current velocity is thus the sum of two analogous components: a component that depends on the slope of the free surface of the sea and remains constant with depth and a component that is induced by the internal pressure field as a result of the disturbance of the water density field. The last component depends on the vertical coordinate. We must agree with this resolution of acting forces and current velocity vector into two such components. But later, Sverdrup concludes that the only component of sea current velocity which can be computed from data on the distribution of sea water temperature and salinity by the dynamic method is the one induced by the disturbance of the density field. In his opinion, the gradient component of current velocity induced by the slope of the sea surface cannot be computed. Sverdrup warns against erroneous conclusions due to confusion between these "different" phenomena.

If we adopt Sverdrup's point of view, we must assume that the gradient current created by the slope of the sea surface penetrates to any depth and has a velocity at the bottom that is of the same order of magnitude as velocity at the surface. The current (induced by the horizontal inhomogeneity of the field of mass, according to Sverdrup) computed by the dynamic method usually decreases with depth and its velocity reaches zero at some horizon. This means that a steady gradient current with a speed of several centimeters and even several tens of centimeters per second must always exist at great ocean depths since the sea surface is practically never horizontal.

From this point of view, it remains unclear why the current induced by the horizontal inhomogeneity of the water density field must decrease with depth. It would seem that the horizontal pressure gradient in a gravitational field created by the slope of the isopycnal surfaces must increase with depth from zero at the free surface to a maximum value at the bottom if the slope of isopycnal surfaces is the same throughout the water. The vertical variation of the corresponding component of current velocity must behave similarly. Hence we can see that Sverdrup's conclusions are not without contradiction.

Having compared the above two views regarding the dynamic method, we can see that they are contrary. According to N. N. Zubov, the dynamic method is universal and can be used to compute all the components of a steady sea current. On the other hand, H. Sverdrup believes that this method can be used to compute only that component of gradient current velocity which is due to the horizontal inhomogeneity of the density field.

To resolve this contradiction we will try to determine the component of sea current velocity that is being computed by the dynamic method from the resultant structure of the water density field. From simplified fluid motion equations, we will derive a dynamic formula in its standard form. As we know, the dynamic method formula is derived from the circulation theorem, but in doing this, a number of very important conclusions for the application of the method are concealed. Such conclusions can become apparent when the formula (1.1) is derived from equations of geostrophic fluid motion. During the process of derivation it will become clear which component of current velocity is being computed by the dynamic method. It is very important to know this in applying the method.*

The component of current velocity normal to the vertical cross section in the steady state is determined from the relationship

$$\Omega v \rho = \frac{\partial p}{\partial x}, \qquad (1.3)$$

without consideration of frictional forces.

---

*The derivation of the formula is not original, but the conclusions drawn from it are.

As the second equation let us use the hydrostatic equation

$$g\rho = \frac{\partial p}{\partial z}. \qquad (1.4)$$

Using the designations adopted earlier, here ρ is the density of sea water; $\Omega = 2\omega \sin \varphi$ is the coriolis parameter; and g is the acceleration of gravity. The system of coordinates is shown in Fig. 1. At the sea surface $z = \zeta(x)$. Assuming that atmospheric pressure is constant, from (1.4) we obtain

$$p - p_\mathrm{a} = g \int_\zeta^z \rho \, dz; \qquad (1.5)$$

$$\frac{\partial p}{\partial x} = g \int_\zeta^z \frac{\partial \rho}{\partial x} dz - g\rho(\zeta) \frac{\partial \zeta}{\partial x}, \qquad (1.6)$$

where p_a is atmospheric pressure; $\rho(\zeta)$ is water density at the sea surface.

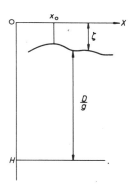

Fig. 1. System of coordinates. $z = \zeta(x)$ —equation of the trace of the free surface of the sea. There is no motion at depth $z = H$.

Equation (1.6) shows that the horizontal pressure gradient in the sea consists of two components. One is due to the horizontal inhomogeneity of the water density field and depends largely on the vertical coordinates, while the other results from the slope of the free surface of the sea and does not change along the vertical. Obviously, we can assume, without introducing a large error, that the lower boundary of the integral in expression (1.6) is equal to zero since $\zeta \ll z$ with actual values of z. But to keep the generality of the arguments we shall not make this substitution.

Substituting (1.6) in (1.3) we find

$$\Omega v = \frac{g}{\rho} \int_{\zeta}^{z} \frac{\partial \rho}{\partial x} dz - g \frac{\rho(\zeta)}{\rho} \frac{\partial \zeta}{\partial x}. \qquad (1.7)$$

Let us now assume that at a certain depth $z = H$ the horizontal pressure gradient and current velocity are zero. This is the same as saying that the "zero" isobaric surface is located at a depth $z = H$. Then from (1.6) or (1.7) it follows that

$$g \int_{\zeta}^{H} \frac{\partial \rho}{\partial x} dz = g\rho(\zeta) \frac{\partial \zeta}{\partial x}. \qquad (1.8)$$

Let us transform the integral in expression (1.8) and give it the form ordinarily used in the dynamic method. To do this, we must replace density ρ by specific volume $\alpha = \frac{1}{\rho}$ and integrate with respect to the variable p (pressure).

$$\int_{\zeta}^{H} \frac{\partial \rho}{\partial x} dz = - \frac{\bar{\rho}}{g} \frac{\partial}{\partial x} \int_{p_a}^{p_H} \alpha\, dp. \qquad (1.9)$$

Here $\bar{\rho}$ is the average density of sea water in the section $\{p_a, p_H\}$ and p_a and p_H are pressure at the sea surface and at horizon $z = H$, respectively. We can simply rearrange integration and differentiation operations because we assume that

$$\frac{\partial p_a}{\partial x} = \frac{\partial p_H}{\partial x} = 0. \tag{1.10}$$

The integral on the right side of relationship (1.9) is the known expression for the dynamic height D of the sea surface. Having combined (1.8) and (1.9), we obtain

$$-\frac{\partial D}{\partial x} = g\frac{\rho(\zeta)}{\bar{\rho}}\frac{\partial \zeta}{\partial x} \tag{1.11}$$

or

$$-\frac{\partial D}{\partial x} = g\frac{\partial \zeta}{\partial x}, \tag{1.12}$$

since $\frac{\rho(\zeta)}{\bar{\rho}} \approx 1$.

Having introduced $z = \zeta$ in expression (1.7) and keeping in mind relationship (1.11), we obtain a formula for computing current velocities at the sea surface $z = \zeta(x)$

$$\Omega v = \frac{\partial D}{\partial x}, \tag{1.13}$$

where

$$D = \int_{p_H}^{p_a} \alpha\, dp. \tag{1.14}$$

Let us integrate equation (1.12) with respect to x:

$$\zeta(x) = \frac{D}{g} + \text{const}. \tag{1.15}$$

When computing current velocity in the sea it is not necessary to know the absolute deviation of the sea surface from the unperturbed level. Therefore, in determining the integration constant let us assume that $\zeta = 0$ with $x = x_0$. Moreover, if $D = D_0$ with $x = x_0$, then, instead of (1.15), we have

$$\zeta(x) = \frac{D_0 - D}{g}. \tag{1.16}$$

Let us discuss the results. Initially, we used the equation of geostrophic fluid motion and obtained a formula for computation by the dynamic method. This means that the current computed by the dynamic method is geostrophic and is associated with the horizontal pressure gradient that is balanced on the rotating earth by the coriolis force which results from motion. As we can see from equation (1.6), the horizontal pressure gradient is expressed by two terms. One is the result of the slope of the sea surface and the other, the result of the horizontal inhomogeneity of the water density field.

When deriving the formula for computation by the dynamic method, can it be assumed that the transverse slope (relative to the current) of the free surface of the sea or any other isobaric surface in the water is created by the same current velocity component computed by the dynamic method? In the author's opinion, this assumption is hardly correct. The steady gradient current computed by the dynamic method cannot cause the inclination of isobaric surfaces, but is itself the result of the slope of these surfaces. A similar phenomenon is encountered in the atmosphere where variations in the distribution of atmospheric pressure cause the formation of gradient wind. The analogy between these two phenomena consists mainly in the fact that they can both be described by geostrophic relationships.

Proceeding from the equations of geostrophic fluid motion (1.3) and (1.4), we obtained, with some transformation, a formula for computation by the dynamic method. Therefore, equations (1.3) and (1.4) as well as all subsequent intermediate expressions relate to the current velocity component computed by the dynamic method from the distribution of water density. As we know, equations of motion are the symbolic expression of the equilibrium of forces acting on a water particle. In our case two forces act on a water particle in the horizontal plane: the horizontal pressure gradient and the coriolis force $\Omega v \rho$. Since the coriolis force is a secondary force derived from motion, we must consider the force inducing motion to be the horizontal pressure gradient inasmuch as other forces were not taken into consideration.

The author cannot agree that the current computed by the dynamic method produces the transverse inclination of isobaric surfaces in the coriolis force field. The horizontal

1. DYNAMIC METHOD OF COMPUTING SEA CURRENTS

pressure gradient is not the result, but the cause of this current velocity component. In other words, the current being computed by the dynamic method is not simply *associated* with the horizontal pressure gradient, but is *induced by the horizontal pressure gradient* (i.e., it is a *gradient current* by nature).

Later, we shall discuss the reasons for the formation of a horizontal pressure gradient in the sea and, consequently, the nature of gradient currents. It is enough to point out here that the dynamic method cannot be used to compute the pure drift component of the velocity of a steady wind-driven current because frictional forces that are not considered in the dynamic method do play a decisive role in the formation of this current velocity component. Even though we know that the pure drift component of wind-current velocity creates and sustains the slope of isobaric surfaces and, in particular, the slope of the sea surface, it cannot be computed from the density field. Therefore, in deriving a formula we cannot fully disregard factors responsible for currents without running the risk of drawing the wrong conclusions.

Let us return to the analysis of equation (1.6). The first term of expression (1.6) becomes zero at the free surface of a baroclinic sea or at some depth in a homogeneous sea. In these cases, motion is induced only by the slope of the sea surface. In a baroclinic sea, where current velocity and the horizontal pressure gradient reach zero at some depth, the first and second terms of formula (1.6) must be generally of opposite sign.* The second term does not depend on the vertical coordinate, while the first term increases in absolute value with depth. This means that with increasing depth the effect of the slope of the sea surface is gradually compensated by the effect of the inhomogeneity of the water density field, while the total horizontal pressure gradient $\frac{\partial p}{\partial x}$ decreases along the vertical. The horizontal pressure gradient and, consequently, the gradient component of current velocity become

---

*It should be remembered that with the adopted method of reading the excess of ζ, the derivative of $\partial \zeta / \partial x$ and the horizontal pressure gradient corresponding to it already have different signs.

zero at some depth. If this is not so, and if both components of the horizontal pressure gradient do not have opposite signs (even on the average), the gradient current will eventually include the entire layer of water and will reach the bottom with a higher velocity than at the sea surface. This conclusion follows directly from equation (1.7).

Expressions (1.7) and (1.8) can be used to formulate the physical meaning of the concept of the "zero" surface. *The "zero" surface is that surface in the sea at which complete mutual compensation of two horizontal components of the pressure gradient takes place: that component due to the slope of the free surface of the sea and that due to the inhomogeneity of the field of mass.*

Relationship (1.8) can be used to estimate approximately the depth of the "zero" surface. To do this, we must assign a value to the gradient component of current velocity at the sea surface (the slope of the free surface of the sea) and to the characteristic (average along the vertical) value of the horizontal sea water density gradient in the region under study. We shall discuss this in more detail in section 4.

In our preceding arguments, we assumed that $z = \zeta(x)$ describes the position of the free surface of the sea. The only requirement was that atmospheric pressure p_a at the sea surface be constant (i.e., that the sea surface be an isobaric surface). If we now repeat all these arguments, assuming that $z = \zeta(x)$ is the equation of an arbitrarily selected isobar within the water rather than the equation for the trace of the sea surface in a vertical section, then the conclusions remain the same. This means that from this point on we can consider $z = \zeta(x)$ as the equation for the trace of any isobaric surface in a vertical section, including the sea surface, and that the formula derived for computation by the dynamic method can be used to compute gradient current velocity at any isobaric surface. Moreover, we could introduce a second equation for geostrophic fluid motion and probably obtain relationships similar to expressions (1.5)-(1.16) that would be valid for the second horizontal component of gradient current velocity in a rectangular system of coordinates XYZ. This makes it possible to speak, therefore, of the sea surface

1. DYNAMIC METHOD OF COMPUTING SEA CURRENTS

and other isobaric surfaces instead of their traces in a vertical section and of the full horizontal pressure gradient. Let us now examine equation (1.16). This relationship demonstrates the unity of the geometric and dynamic relief of an isobaric surface. The dynamic relief of the sea surface or any other isobaric surface is the same as the geometric relief, but expressed on another scale. A dynamic isobaric surface chart, constructed relative to an accurately selected "zero" surface, describes the deviation of the isobaric surface from its undisturbed position (i.e., a horizontal plane), with a precision to the value of the constant. To reduce dynamic heights D to $\zeta(x, y)$ deviations (the latter are measured in geometric units), it is sufficient to divide dynamic heights by the acceleration of gravity.

When the reference surface is not the "zero" surface (i.e., when current velocity at the reference surface is not zero), a dynamic chart describes the relief of the isobaric surface with a precision to the values of the deviation of the isobaric reference surface from its position in the unperturbed state. Then

$$\frac{D_0 - D}{g} = \zeta(x, y) - \zeta^*(x, y), \qquad (1.16)$$

where $\zeta(x,y)$ and $\zeta^*(x,y)$ represent respectively the deviations of the isobaric surface at which current is computed and of the isobaric reference surface from the horizontal plane.

Equation (1.16) shows H. Sverdrup's conclusion to be baseless that the density field can be used only to compute the relative current and not a tidal current. In fact, the current being computed by the dynamic method is induced by the horizontal pressure gradient which, as pointed out earlier, is the result of both the slope of the sea surface and the inhomogeneity of the water density field in the layer from the surface to the given isobaric surface.

At first glance there seems to be a contradiction. We know that the relief of an isobaric surface is computed by the dynamic method from the density distribution in the layer of water bounded on one side by the considered isobaric surface and on the other side by the "zero" surface. In other words, current velocity is computed from the distribution of density in the underlying water layer. This contradiction disappears if we remember the above definition of the "zero" surface as that surface at which complete mutual compensation of the two horizontal components of the horizontal pressure gradient takes place. From this follows a very important feature of the stationary distribution of density in a baroclinic sea which makes it possible to compute steady gradient currents by the dynamic method. *The horizontal pressure gradient produced in a baroclinic sea by the slope of the sea surface and the inhomogeneity of the density field in the layer of water overlying the isobaric surface under consideration is equal in absolute value to the horizontal pressure gradient produced by the inhomogeneity of water density in the underlying layer (from the isobaric surface under consideration to the "zero" surface).* In the opposite case, i.e., when the formulated assumption is not fulfilled, the surface selected as the "zero" surface is apparently not the actual "zero" surface. If compensation of the two horizontal pressure gradients above is not achieved when the reference surface approaches bottom, then the geostrophic gradient current penetrates to the bottom.

In conclusion, let us mention again that the gradient current is computed by the dynamic method from the distribution of sea water density. To determine its nature in more detail we must examine the processes that lead to distortion of the free surface and to the horizontal inhomogeneity of the field of mass of the sea. These include, first, the effect of wind on the sea surface as well as unequal climatic conditions over a sea basin. The following sections take up these problems.

In our opinion, the reason for contradiction concerning the nature of the component of current velocity being computed is its obvious geometrical interpretation which considers the variation of velocity circulation along a closed contour to be numerically equal to the number of single Bjerknes' solenoids bounded by this contour. Some investigators believe that the method is universal, assuming that the density field on a rotating earth must adjust in an equal manner to any current. Others narrow the possibilities of the dynamic method and assume that only the velocity of a current induced by the horizontal inhomogeneity of the field of mass can be computed from the distribution of density. As we can see, these two opinions differ only in the selection of the causes and effects of the phenomenon. According to the first point of view, the horizontal inhomogeneity of the field of mass is the result of flow, while the second considers it to be its cause. However, both these viewpoints are wrong because they deny the mutual adjustment of the density and velocity fields in the sea.

2. WIND-DRIVEN CURRENTS IN AN INFINITE CHANNEL. PURE DRIFT AND GRADIENT COMPONENTS OF THE VELOCITY OF A WIND-DRIVEN CURRENT. ROLE OF FRICTION IN A GRADIENT CURRENT. MECHANISM OF THE MUTUAL ADJUSTMENT OF THE DENSITY AND CURRENT VELOCITY FIELDS IN THE SEA

From the theory of wind-induced ocean currents we know that the effect of wind on the sea surface induces the forward movement of water particles, produces the slope of the free surface of the sea, and results in the readjustment of the density structure. Computations show that the chief cause of sea currents is wind. Therefore, to understand better the theoretical principles of the dynamic method one could trace the mechanism of the formation of wind-driven currents and the main patterns of the vertical distribution of current velocity using some idealized model. This is possible, since one of the components of a steady wind-driven current, namely, the gradient component, can be computed from the resultant structure of the water density field.

The strict theoretical solution of the problem of the development of wind-driven currents in a sea involves difficult mathematics. There are only a few papers that solved this problem more or less rigorously. Here we shall discuss the major results of the investigations of P. S. Lineikin (1955). This will permit us to make a number of general conclusions concerning the mechanism of the formation of wind-driven currents in an inhomogeneous sea, to trace the effect of individual factors on their vertical velocity distribution pattern, and to give a comparative estimate of these factors. As we can readily see, this has a direct bearing on the application of the dynamic method.

Following P. S. Lineikin, let us examine an unlimited, infinitely deep channel with vertical shore boundaries that extend in the direction of the y-axis. The origin of coordinates is on one of the shore boundaries of the channel so that on the shore boundaries $x = 0$ and L. The z-axis is directed vertically downward.

To solve the problem we write the following closed system of equations:

$$\rho \frac{\partial u}{\partial t} - 2\omega_z \rho v = -\frac{\partial p}{\partial x} + \mu_x \Delta u + \frac{\partial}{\partial z}\left(\mu_z \frac{\partial u}{\partial z}\right); \quad (1.17)$$

$$\rho \frac{\partial v}{\partial t} + 2\omega_z \rho u = -\frac{\partial p}{\partial y} + \mu_x \Delta v + \frac{\partial}{\partial z}\left(\mu_z \frac{\partial v}{\partial z}\right); \quad (1.18)$$

$$g\rho = \frac{\partial p}{\partial z}; \quad (1.19)$$

$$\frac{\partial \rho}{\partial t} + \frac{\partial \rho u}{\partial x} + \frac{\partial \rho v}{\partial y} + \frac{\partial \rho w}{\partial z} = 0; \quad (1.20)$$

$$\frac{d\rho}{dt} = \frac{k_x}{\rho_0}\Delta \rho + \frac{1}{\rho_0}\frac{\partial}{\partial z}\left(k_z \frac{\partial \rho}{\partial z}\right); \quad (1.21)$$

$$\Delta \equiv \frac{\partial^2}{\partial x^2} + \frac{\partial^2}{\partial y^2}.$$

This system of equations is solved with the following boundary conditions.

At the initial moment:

$$t = 0, \quad u = v = w = 0, \quad \rho = \rho_*(z).$$

At the sea surface:

$$z = \zeta; \quad p = p_a; \quad \mu_z \frac{\partial u}{\partial z} = -T_x; \quad \mu_z \frac{\partial v}{\partial z} = -T_y.$$

With $z \to \infty$

$$u = v = w \to 0; \quad (\rho - \rho_*) = (p - p_*) \to 0.$$

2. WIND-DRIVEN CURRENTS IN AN INFINITE CHANNEL

On the shores of the channel:

$$x = 0 \text{ and } L; \quad u = v = 0; \quad \frac{\partial \rho}{\partial n} = 0.$$

The sea surface relief is determined from the relationship

$$\frac{\partial \zeta}{\partial t} = w_{z=0}.$$

The following expressions are used to reduce the system of initial equations to a dimensionless form:

$$x = L\bar{x}; \quad y = L\bar{y}; \quad z = h\bar{z}; \quad t = \tau\bar{t}; \quad \zeta = h\bar{\zeta};$$

$$\rho = \rho_0\bar{\rho}; \quad u = u_0\bar{u}; \quad v = u_0\bar{v}; \quad w = w_0\bar{w}; \quad T_x = T_0 X;$$

$$T_y = T_0 Y; \quad h = L\sqrt{\frac{\mu_z}{\mu_x}}; \quad \tau = \frac{\sqrt{\mu_x \mu_z}}{T_0};$$

$$u_0 = \frac{LT_0}{\sqrt{\mu_x \mu_z}}; \quad w_0 = \frac{LT_0}{\mu_x}; \quad c = \frac{2\omega_z \sqrt{\mu_x \mu_z}}{T_0};$$

$$a = \frac{g\mu_z \sqrt{\mu_x \mu_z}}{LT_0^2}; \quad \gamma = bL\sqrt{\frac{\mu_z}{\mu_x}}; \quad \lambda = \frac{\mu_x \sqrt{\mu_x \mu_z}}{\rho_0 T_0 L^2}.$$

The following designations are used in the above expressions: ρ for sea water density; ρ_* for sea water density at the initial moment; ρ_0 for water density at the sea surface with $t = 0$; u, v, w for the components of current velocity along the coordinate axes x, y, z, respectively; t for time; $\omega_z = \omega \sin \varphi$ for the vertical component of the angular speed of rotation of the earth; p for pressure; p_* for pressure at the initial moment; $p_a = \text{const}$ for atmospheric pressure; μ_x and μ_z for the coefficients of the horizontal and vertical exchange of the amount of movement, respectively; g for the acceleration of gravity; k_x and k_z for the coefficients of density diffusion in the horizontal and vertical directions, respectively [henceforth we shall assume that $k_x = \mu_x$, $k_z = \varepsilon\mu_z$ ($\varepsilon = 0.1$)]; T_x and T_y for the component of tangential wind stress along the x and y axes; T_0 for the characteristic value of tangential wind stress; and n for the direction of the normal to the boundary formed by the shoreline.

It is assumed that the turbulent exchange coefficient is constant. Sea water density at the initial moment is given by

$$\rho_* = \rho_0 (1 + bz),$$

that varies linearly, where $\rho_0 b \approx b$ is the average vertical density gradient. Moreover, nonlinear terms are not considered in the equations of motion and density diffusion. Boundary conditions with the free surface $z = \zeta$ are transferred to plane $z = 0$. The compressibility of sea water is not considered.

In this case, the solution of the equations, taking into account the above formulated boundary conditions for a steady-state current (if we limit ourselves to the first terms of expansion), has the form:

$$\frac{\bar{u}}{\sin \pi x} = \frac{\sqrt{\bar{\lambda}}}{r} Re^{-\frac{\alpha \bar{z}}{\sqrt{\bar{\lambda}}}} \cos\left(\varphi - \psi - \frac{\beta \bar{z}}{\sqrt{\bar{\lambda}}}\right) - \frac{\pi \lambda \sqrt{\bar{\gamma a}}}{r^4} R \cos(\varphi - 2\psi) e^{-\frac{\pi \sqrt{\bar{\gamma a}}}{r^2} \bar{z}}; \qquad (1.22)$$

$$\frac{\bar{v}}{\sin \pi x} = \frac{\sqrt{\bar{\lambda}}}{r} Re^{-\frac{\alpha \bar{z}}{\sqrt{\bar{\lambda}}}} \sin\left(\varphi - \psi - \frac{\beta \bar{z}}{\sqrt{\bar{\lambda}}}\right) + \frac{\pi \lambda \sqrt{\bar{\gamma a}}}{r^4} R \sin(\varphi - 2\psi) e^{-\frac{\pi \sqrt{\bar{\gamma a}}}{r^2} \bar{z}}; \qquad (1.23)$$

$$\frac{\bar{w}}{\cos \pi x} = \frac{\pi \lambda}{r^2} Re^{-\frac{\alpha \bar{z}}{\sqrt{\bar{\lambda}}}} \cos\left(\varphi - 2\psi - \frac{\beta \bar{z}}{\sqrt{\bar{\lambda}}}\right) - \frac{\pi \lambda}{r^2} Re^{-\frac{\pi \sqrt{\bar{\gamma a}}}{r^2} \bar{z}} \cos(\varphi - 2\psi); \qquad (1.24)$$

$$\frac{\delta}{\cos \pi x} = \frac{\gamma \pi \lambda}{\sigma r^2} Re^{-\frac{\alpha \bar{z}}{\sqrt{\bar{\lambda}}}} \cos\left(\varphi - 2\psi - \eta - \frac{\beta \bar{z}}{\sqrt{\bar{\lambda}}}\right) + \frac{\gamma R}{\pi r^2} e^{-\frac{\pi \sqrt{\bar{\gamma a}}}{r^2} \bar{z}} \cos(\varphi - 2\psi); \qquad (1.25)$$

$$\frac{\bar{\zeta}}{\cos \pi x} = -\frac{\pi \gamma \lambda^{3/2}}{r^3 \sigma} R \cos(\varphi - 3\psi + \eta) + \frac{\sqrt{\bar{\gamma}}}{\pi^2 \sqrt{a}} R \cos(\varphi - 2\psi) + \frac{\varepsilon \gamma \sqrt{\bar{\lambda}}}{\pi \sigma r} R \cos(\varphi - \psi + \eta). \qquad (1.26)$$

2. WIND-DRIVEN CURRENTS IN AN INFINITE CHANNEL

In these expressions

$$R = \sqrt{X^2 + Y^2}; \quad \varphi = \arctan \frac{Y}{X}; \quad \psi = \frac{1}{2} \arctan \frac{c}{\lambda \pi^2};$$

$$r = \sqrt{\lambda^2 \pi^4 + c^2}; \quad \alpha = r \cos \psi; \quad \beta = r \sin \psi;$$

$$\eta = \arctan \frac{\varepsilon c}{\lambda \pi^2 (1 - \varepsilon)};$$

$$\sigma = \sqrt{\lambda^2 \pi^4 (1 - \varepsilon)^2 + \varepsilon^2 c^2},$$

$\delta = (\rho - 1 - \gamma z)$ is the disturbance of density.

The components of the velocity of a wind-driven current are represented in formulas (1.22)-(1.24) as the sum of two monomials. It is easy to show that the first terms define the pure drift component of the current velocity that is directly produced by the entraining action of the wind. The second terms define the current velocity component that results from the secondary effect of the wind. P. S. Lineikin calls it the *convective-gradient* current velocity component, thus emphasizing that it depends on both the slope of the free sea surface and on the readjustment of the density structure of an inhomogeneous sea. But we shall retain the accepted terminology and call it the gradient component since it is induced by the horizontal pressure gradient.

If we assume, as it is done in Ekman's classical theory of wind-driven currents, that the wind direction coincides with the positive direction of the y-axis, then the horizontal component of a pure wind-driven current at the axis of the channel will be described by the following expressions:

$$\bar{u} = \frac{\sqrt{\lambda}}{r} Y e^{-\frac{\alpha \bar{z}}{\sqrt{\lambda}}} \cos\left(\psi + \frac{\beta \bar{z}}{\sqrt{\lambda}}\right);$$

$$\bar{v} = \frac{\sqrt{\lambda}}{r} Y e^{-\frac{\alpha \bar{z}}{\sqrt{\lambda}}} \sin\left(\psi + \frac{\beta \bar{z}}{\sqrt{\lambda}}\right),$$

(1.27)

that outwardly resemble the known formulas of Ekman (1905).

The absolute current velocity at the sea surface in expressions (1.27) is

$$U_{z=0} = \frac{\sqrt{\bar\lambda}}{r} Y.$$

Substituting here the values of the quantities included and neglecting $\lambda^2 \pi^4$ as compared to c^2, we obtain a formula that exactly corresponds to Ekman's formula

$$U_{z=0} = \frac{T}{\sqrt{2\omega_z \rho_0 \mu_z}}. \qquad (1.28)$$

From the solution it follows that the pure drift component of current velocity at the sea surface deviates from the wind direction in the Northern Hemisphere by an angle ψ that is determined from the relationship

$$\psi = \frac{1}{2} \arctan \frac{2\omega_z \rho_0 L^2}{\pi^2 \mu_x}. \qquad (1.29)$$

This angle does not depend on wind direction, but varies with channel width, horizontal turbulent friction coefficient, and geographic latitude.

Using the "four-thirds" law of Richarson and Obukhov (Defant, 1954), $\mu_x = 0.05 L^{4/3}$, and having selected L (channel width) as the characteristic scale, we obtain an expression for ψ as a function of L and φ. This functional relationship is presented in Fig. 2. Only in the case of an infinitely wide channel does the angle of deviation of the pure drift component of the velocity of a wind-driven current from the wind direction at the sea surface amount to 45°. Actually, it differs little from this value in all other cases.

The pure drift component of current velocity decreases exponentially with depth; the attenuation constant

$$\frac{\alpha \bar z}{\sqrt{\bar\lambda}} = \sqrt{\frac{2\omega_z \rho_0}{\mu_z}} \cdot z \cos \psi \qquad (1.30)$$

corresponds exactly to Ekman's dimensionless value az with $\psi = 45°$.

The variation in the direction of the vector of a pure wind-driven current with depth, given by expression

$$\frac{\beta \bar z}{\sqrt{\bar\lambda}} = \sqrt{\frac{2\omega_z \rho_0}{\mu_z}} z \sin \psi. \qquad (1.31)$$

2. WIND-DRIVEN CURRENTS IN AN INFINITE CHANNEL

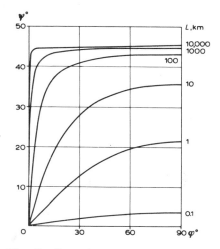

Fig. 2. Dependence of angle ψ on the geographical latitude of the area and the width of the channel.

is also exactly identical with Ekman's law of the vertical variation in the direction of a wind-driven current with $\psi = 45°$.

From this it follows that the pure drift component of current velocity in a channel agrees with the results of Ekman's theory with an accuracy to a small value. By small value we mean the value of $\lambda^2 \pi^4$ as compared to c^2. If we use the law of the "four-thirds" for μ_x, then their relationship

$$\frac{c^2}{\lambda^2 \pi^4} = 4\omega_z^2 L^{4/3} \qquad (1.32)$$

shows that the value being neglected is actually small with large φ and L. This assumption is invalid only in the immediate vicinity of the equator or when the width of the channel is very small.

The gradient component of the velocity of a wind-driven current is described by the second terms in formulas (1.22)-(1.24). It decreases exponentially with depth, but attenuates much more slowly with depth than does the pure drift component of current velocity. The intensity of the vertical variation of the gradient component of a wind-driven current depends on sea water stratification. Its direction does not change with depth. In the Northern Hemisphere the wind-driven gradient current deflects to the left of the geostrophic current by an angle of $90°$ (2ψ). The dependence of angle ψ on geographic latitude and on the width of the channel has been shown earlier. Since it is rare for ψ to be an exact $45°$, the gradient component of the velocity of a wind-driven current always forms a certain angle with the direction of the geostrophic current.

Figure 3 shows vector diagrams of the pure drift and gradient components of current velocity on the axis of a channel according to Lineikin. It is assumed that a 10-m/sec wind is directed along the channel, $\varphi = 45°$, $\mu_x = 10^8$ cgs, $\mu_z = 10^2$ cgs, and $L = 10^7$ cm. Figure 4 shows the vertical distribution of the gradient component of current velocity under the same conditions.

The deviation of water density δ from its unperturbed value as a result of the action of wind at a particular point

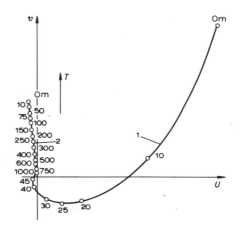

Fig. 3. Hodographs of the pure drift (1) and gradient (2) components of the velocity of a steady wind-driven current at the axis of the channel (according to P. S. Lineikin). The arrow shows wind direction.

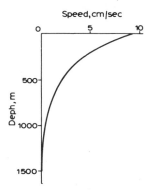

Fig. 4. Vertical distribution of the gradient component of velocity of the steady wind-driven current in the channel (according to P. S. Lineikin).

is expressed in the solution by three terms that describe the influence of pure drift, of the gradient component of current velocity, and of turbulent exchange (diffusion) on the distribution of sea water density. The turbulent term rapidly attenuates with depth so that the disturbance of the density field beyond the friction layer (mixed surface layer) is entirely determined by the gradient component of current velocity. The deviation of the sea surface ζ from the position of the unperturbed surface is also the result of the combined action of these three components.

Lineikin's solution indicates that the gradient component of current velocity does not coincide in direction with the geostrophic current. This makes it possible to represent in general outline the mechanism of the mutual adjustment of the sea water density and current velocity fields.

A wind acting on the sea surface induces the pure drift component of current velocity. This velocity component penetrates to a relatively small depth, of the order of several tens of meters. As we know, the full pure wind-driven current is directed to the right of the wind in the Northern Hemisphere. The transport of water to the right and the effect of shores and irregularities of the wind field create regions where the water rises and sinks and where the sea surface slopes. As the latter takes place, a horizontal pressure gradient forms that induces the gradient component of velocity in the entire layer of water. Owing to frictional forces, the gradient current is not geostrophic, but deviates from the direction of the latter by some angle. As a result, transverse circulation forms in the channel and, in a general case, in any vertical section of a sea basin. This leads to the readjustment of the density structure in the sea. The concentration of lighter water at the right of the current and of heavier water at the left is achieved both by the transport of water by the pure wind-driven current in the upper layer and transverse circulation in the gradient current. The transverse circulation of a gradient current in a channel is shown schematically in Fig. 5. The wind is assumed to be directed into the diagram.

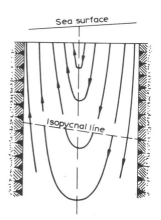

Fig. 5. Transverse gradient circulation in the channel (according to P. S. Lineikin).

The readjustment of the density field generates the second component of the horizontal pressure gradient that has a sign opposite to that of the pressure gradient which is produced by the slope of the free surface of the sea. The total horizontal pressure gradient and,

2. WIND-DRIVEN CURRENTS IN AN INFINITE CHANNEL

consequently, the magnitude of the gradient component of current velocity decrease with depth, indefinitely approaching zero. This leads to the conclusion that the gradient component of the velocity of a steady wind-driven current is a current that can be computed by the dynamic method from the resultant structure of the water density field.

Using the computed density and current velocity fields, P. S. Lineikin compared the geostrophic current velocity v_g, computed from the density field by the dynamic method, with the gradient component of current velocity obtained with consideration of frictional forces. If we limit ourselves to the first term of the expansion, their ratio can be estimated from the following expression:

$$\frac{v_g}{v} = 1 + \frac{\pi^4 \mu_x^2}{4\omega^2 \rho^2 \sin^2 \varphi \cdot L^4} \qquad (1.33)$$

or

$$\frac{v_g}{v} = 1 + \frac{0.05 \pi^4}{4\omega^2 \rho^2 L^{4/3} \sin^2 \varphi}, \qquad (1.34)$$

since $\mu_x = 0.05 L^{4/3}$.

In the example under consideration ($L = 10^7$ cm, $\varphi = 45°$) this ratio is 1.01 (i.e., geostrophic velocity exceeds the gradient component of current velocity computed with consideration of the frictional forces by only 1%). The error in the full current computed from the geostrophic current is smaller by a factor of approximately 3.

This result proves the validity of neglecting frictional forces on a steady gradient current and serves as a theoretical justification, as it were, of the dynamic method. The influence of frictional forces on the gradient component of current velocity in the steady state is exceedingly small. The current can be considered with a sufficient degree of accuracy as geostrophic. But the effect of friction becomes decisive in the mutual adjustment of the density and current velocity fields. Their mutual adjustment is impossible without friction since the transverse and vertical components of velocity that produce the redistribution of the density field result from friction.

The dynamic method makes it possible to compute current velocity from the observed water density distribution. The stationary distribution of density in a sea already contains the effect of frictional forces. Therefore, we can say that although the dynamic method neglects the influence of frictional forces on current velocity, it takes their effect into account in the final analysis through the observed density distribution. We mean here the effect that frictional forces had on the vertical distribution of the gradient component of current velocity when it became developed and stabilized.

3. SOME CONSIDERATIONS CONCERNING CONVECTIVE COMPONENTS OF CURRENT VELOCITY IN THE SEA. CURRENTS INDUCED BY A DIFFERENCE IN CLIMATIC CONDITIONS OVER THE OCEAN AND BY THE TRANSPORT OF HEAT AND SALT IN A WIND-DRIVEN CURRENT

In examining the principles of the dynamic method of computing ocean currents, we showed that the computed current velocity is determined both by the slope of the free surface of the sea and by the horizontal inhomogeneity of the water density field. One cause for the formation of the slope of the sea surface and of horizontal density gradients is the effect of wind on the sea surface. Let us now discuss the other reasons for the formation of gradient currents in a sea, namely, uneven differences in climate over the ocean and mass transport by the wind current.

Convective Currents Resulting from Differences
in Climatic Regimes and their Effects on
the Sea Surface

Oceans, extending from one pole to the other, occupy a tremendous space on the earth's surface. Various sections of the ocean are located in different climatic zones. The amount of solar radiation on a unit area of water surface varies with latitude and the amounts of precipitation and evaporation from the sea surface differ in various parts of the ocean. All these factors affect sea water density to some degree and produce horizontal inhomogeneity.

3. CONVECTIVE COMPONENTS OF CURRENT VELOCITY

If we select a horizontal surface at some depth, then the pressure of the overlying water column on this surface will differ in various parts of the ocean as a result of horizontal variations in density. Consequently, the heterogeneity of climatic conditions over the ocean (heating and cooling of surface water, precipitation and evaporation) must lead to the formation of horizontal pressure gradients and, consequently, to convective currents tending to balance this inhomogeneity.*

A number of studies by domestic and foreign scientists are devoted to climate-induced convective circulation in the ocean. Thus, Ekman (1926) and Goldsbrough (1933) have analyzed circulation produced by the uneven distribution of precipitation and evaporation over the sea surface. Having made his computations on the basis of a schematic model of the North Atlantic, Goldsbrough came to the conclusion that evaporation and precipitation generate a system of currents that, in general, resembles the observed system, but the order of magnitude of velocity proved to be too small. V. V. Shuleikin (1945), Stommel (1950), Takano (1955) and others have studied currents resulting from the difference in heating of various parts of the sea surface. The results of these studies show that uneven heating of the sea surface generates convective currents, but that their velocities under actual conditions are insignificant.

Let us examine the mechanism of the formation of convective currents in its most elementary form. This must be done to show that the current velocity component can be computed by the dynamic method from the distribution of water density. Let us assume, for simplicity, that the heating and cooling of water are not accompanied by evaporation and that precipitation and evaporation do not change water density. These assumptions are not strictly true, but they simplify the phenomenon without distorting it to any degree.

---

*Concerning climatic conditions over the ocean we mean the combination of all weather characteristics with the exception of the mechanical influence of wind on the sea surface. Wind is considered only where it affects the intensity of evaporation of the sea surface and heat exchange. This stipulation is necessary because, strictly speaking, wind is one of the most important components of climate.

According to the above, then, the heating and cooling of a water column takes place without a change in its mass. Therefore, pressure remains constant at the depth of penetration of the thermal influence of the atmosphere. The heating (cooling) of water leads to a decrease (increase) of its density and at the same time the water column undergoes thermal expansion (contraction). This means that as a result of uneven distribution of heating, the water surface slopes and a horizontal pressure gradient and a gradient component of current velocity are generated. The horizontal pressure gradient produced by the slope of the sea surface is compensated with depth by the inhomogeneity of the density field. This component of convective current velocity is greatest at the surface, decreases with depth, and reaches zero at the lowest depth which is affected by the thermal influence of the atmosphere.

Since a convective current generated by the uneven distribution of heating of the sea surface is a gradient current by nature and thus depends on the slope of the sea surface and on the inhomogeneity of the field of mass, it can be computed by the dynamic method in a stationary case. Here the "zero" surface must be the lower boundary of the thermally mixed water layer. The influence of frictional forces on this component of the convective current is apparently as negligible as it is in the case of the gradient component of the velocity of a steady wind-driven current.

Since, according to the assumption, evaporation and precipitation do not change water density but merely remove or add a mass of water, their uneven spatial distribution over the water surface produces a slope on the free sea surface, a horizontal pressure gradient, and a gradient current. We can readily see the obvious analogy between the deformation of the sea surface under the influence of wind and its slope resulting from the uneven distribution of precipitation and evaporation. Redistribution of water occurs in both cases: a rise in level in one part of the sea, and a depression in the other. The mutual adjustment of the water density and gradient current velocity fields must evidently occur in the same manner in both cases through transverse circulation generated by frictional forces. In the steady state, this convective component of current velocity is similar to the gradient component of a wind-driven current and can be computed by the dynamic method from the distribution of water density.

Component of Current Velocity Generated by the Transport of Heat and Salt by a Wind-Driven Current

From the theory of wind-driven ocean currents it follows that wind is the main motive force of horizontal water circulation in the oceans and of such strong and steady currents as the Gulf Stream, the Labrador, the East Greenland, the North Atlantic, the Kuroshio, the Oyashio, and other currents. These currents transport large masses of water from one part of the ocean into the other together with thermal energy and a mass of salt. If climatic conditions over the ocean are too homogeneous to create large horizontal water density gradients in its upper layer, a considerable horizontal inhomogeneity of water density of the advective type may arise over a small area owing to the transport of heat and salt by wind-driven currents. Especially large horizontal gradients must occur in regions where wind-driven currents carrying different types of water meet (forming a discontinuity or front), such as the Gulf Stream and the Labrador current, the Kuroshio and the Oyashio, etc. The presence of large horizontal density gradients caused by mass transport in currents must evidently change to some degree the pattern of the vertical distribution of the wind-driven current. This is equivalent to the formation of an additional current velocity component that we shall call a convective current of advective origin.

This kind of current occupies an intermediate place, as it were, between currents generated by mechanical and nonmechanical factors. In fact, the movement of such a current is a nonmechanical factor, i.e., an inhomogeneity of the density field, but at the same time inhomogeneity of the density field is created by the transport of heat and salt in a wind-driven current, i.e., a mechanical factor. Consequently, if the gradient component of the velocity of a wind-driven current is the result of a secondary wind effect, the formation of the current velocity component under discussion can be considered as the result of a third wind effect, i.e., mass transport in a wind-driven current.

Strictly speaking, there is no current generated by mass transport in a pure form. It is unthinkable outside of a wind-driven or other current. Here we must note that resolution of the current velocity vector into components is a conditional operation that is performed only for the convenience of analysis and is permissible only within certain limits. Actually, there is a single current in the sea and not a combination of components. In the case under consideration, mass transport by a wind-driven current creates conditions at the juncture of two currents which carry water of different properties that disturb the pattern of the vertical velocity variation of the normal wind-driven current. Since this vertical variation depends on the additional horizontal density gradient that is created by mass transport, it is convenient to consider an additional component of current velocity, i.e., the convective component of velocity of advective origin. This component of current velocity depends only on the horizontal inhomogeneity of the density field. It is equal to zero at the sea surface and approaches a limiting value with increasing depth. Let us assume that the existence of this current velocity component does not change the slope of the sea surface, since we examine a stationary case and neglect frictional forces. The error is evidently small here, because we assume that the slope of the sea surface created by wind is steep.

To estimate approximately the magnitude of this convective component of current velocity, we use the usual dynamic method and compute a hypothetical example of a "junction," i.e., formation of a discontinuity or front, between a cold and a warm current. Let the vertical water temperature distribution be known at two locations, one in the cold and one in the warm water masses. Water temperature at the sea surface is $5°$ and $10°C$, respectively. Then it decreases uniformly with depth and reaches $2°C$ at the 2000-m level in both water columns. Let us assume for the sake of simplicity that water is isohaline of a

3. CONVECTIVE COMPONENTS OF CURRENT VELOCITY

salinity of 34.50°/oo, the horizontal distance between the water columns is 200 km, and that the average geographic latitude is 50°. Then we shall obtain the following velocities for a convective current of the advective type with a vertical temperature distribution as shown in Fig. 6 (Table 1).

Table 1

z, m	0	250	500	750	1000	1250	1500	1750	2000
v, cm/sec	0	12.4	18.5	20.9	21.8	22.4	22.6	22.7	22.8

The direction of this current is such that higher densities are to the right and lower densities to the left of the current (in the Northern Hemisphere).

When interacting with a wind-driven current, the convective component of current velocity will either intensify or weaken it. This interaction will be examined in detail in the following section. Let us only mention here that it is not always equal to a simple addition of vectors.

In this example we simplified the phenomenon by assuming that density is a function of water temperature

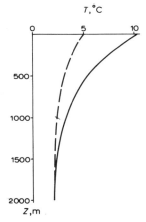

Fig. 6. Vertical water temperature distribution in the warm (solid line) and cold (dashed line) water masses.

only. Under actual conditions warmer water is usually of higher salinity and colder water of lower salinity. Consideration of salt transport by the wind-driven current in this case would have led to a decrease in the computed current velocity, but its general vertical distribution pattern would have remained the same.

4. BAROCLINIC LAYER OF THE SEA. DEPENDENCE OF THE DEPTH OF THE LOWER BOUNDARY OF A CURRENT ON WATER STRATIFICATION AND ON THE WIND FIELD OVER THE SEA

The concept of the baroclinic layer has become firmly established in oceanography as a result of its relation to the theory of wind-driven currents of V. B. Shtokman's method of full currents. Shtokman has established a relationship between the wind field over the sea and total water transport along the vertical. The baroclinic layer is the layer of sea water extending from the surface to some depth, which is under the influence of the wind-induced current. Current velocity and tangential stress must be zero at the lower boundary of this layer. A wind-driven current does not propagate beyond the baroclinic layer.

A fluid is considered to be baroclinic in hydromechanics if its density depends not only on pressure but also on other factors. Consequently, sea water is a baroclinic fluid and the distinction of a baroclinic layer in it is conditional, if the following is considered. As we know, isobaric and isopycnal surfaces in the sea coincide with equipotential surfaces in the absence of movement, i.e., they do not intersect, and this property is inherent to a barotropic fluid. Therefore, sea water at rest has the characteristics of a barotropic fluid, its potentially baroclinic character being demonstrated only in motion. This peculiarity makes it possible to consider moving sea water as baroclinic and a stationary layer of sea water as a barotropic fluid in the first approximation. In this case the density field should be disturbed only within the baroclinic layer. Its lower boundary should coincide with the depth to which

4. BAROCLINIC LAYER OF THE SEA

the current penetrates and isobaric and isopycnal surfaces should be horizontal beyond the baroclinic layer.

If it were possible to observe only a steady wind-driven current at sea, it would be very simple to determine the depth of its penetration (lower boundary of the baroclinic layer of the wind-driven current). To do this, it would be sufficient to analyze the observed vertical sea water density distribution and to determine the depth below which isopycnal surfaces become horizontal. But it is never possible to contemplate a particular phenomenon separately from other phenomena under natural conditions. Sea water particles are not only affected by forces generated by wind. Moreover, various unstable disturbances are always superimposed on a stationary current. This probably explains the absence of horizontal isopycnal lines on hydrological sections.

Let us now examine the dependence of the thickness of the baroclinic layer on various factors and determine the main causes for variations in the depth of its lower boundary. It is evident that the lower boundary of the baroclinic layer and the "zero" surface are the same concepts in computing the gradient component of the velocity of a wind-driven current by the dynamic method. There is no need to clarify this statement further.

Let us rewrite expression (1.8)

$$g \int_{\zeta}^{H} \frac{\partial \rho}{\partial x} dz = g\rho(\zeta) \frac{\partial \zeta}{\partial x}. \qquad (1.8)$$

This relationship formulates the condition for the existence of the lower boundary of the baroclinic layer. The slope of the sea surface created by the gradient current penetrates to a depth H where the horizontal pressure gradient produced by the slope of the sea surface is fully compensated by the readjustment of the density field. If the slope of the sea surface $\frac{\partial \zeta}{\partial x}$ is known and there are data on the density distribution, we can estimate the thickness of the baroclinic layer using equation (1.8). But the magnitude of slope of the free surface is too small to be measured directly and its computation under actual conditions is very complicated. Therefore, expression (1.8) is unsuitable for practical computations. It is used to estimate the potential for a development of a baroclinic layer in the sea. This is accomplished by assigning an order of magnitude to the slope of the sea surface and to the horizontal density gradient, and the approximate thickness of the baroclinic layer is computed from the formula

$$\Omega v \rho = g \rho \frac{\partial \zeta}{\partial x} = g H \left(\frac{\partial \rho}{\partial x}\right)_{\text{cp}} \tag{1.8'}$$

(The only unknown here is depth H). Thus, for example, the sea surface slope is of the order of 10^{-6} and the thickness of the baroclinic layer, 1000 m, with a gradient current velocity of 10 cm/sec at the surface and with an average horizontal density gradient of 10^{-5} per 10 km.

Even such elementary expression as relationship (1.8) makes it possible to draw a number of general conclusions concerning the dependence of the depth of penetration of the current on various factors.

The slope of the free sea surface is determined by the size and configuration of the basin and the wind field over the sea. Hence, the thickness of the baroclinic layer at a given point depends on the geographical and synoptic characteristics of the basin and is not directly associated with wind over this point. Moreover, the vertical stratification of sea water has a significant effect on the depth of penetration of the current because the horizontal density gradient is associated with the vertical density gradient through the angle of inclination of the isopycnal surfaces. From geometrical considerations it follows that

$$\frac{\partial \rho}{\partial x} = \frac{\partial \rho}{\partial z} \operatorname{tg} j, \tag{1.35}$$

where j is the angle of inclination of the isopycnal surface.

Expression (1.35) shows that the horizontal density gradient and, consequently, the rate of attentuation of current velocity with depth depend on the slope of isopycnal surfaces and on the vertical density gradient. In deep layers, where the vertical density gradient is small, even steeply sloping isopycnal surfaces do not produce a significant change in the horizontal components of current velocity along the vertical. In the upper layer the gradient component of current velocity changes rapidly with depth and with small inclinations of isopycnal surfaces because $\frac{\partial \rho}{\partial z}$ is high. Consequently, the depth of penetration of a wind-driven current is inversely proportional to the vertical density gradient: the smaller is $\frac{\partial \rho}{\partial z}$ the greater will be the thickness of the baroclinic layer.

V. B. Shtokman (1953a) was the first to estimate qualitatively the thickness of the baroclinic layer in the sea. If we assume that the thickness of the baroclinic layer in the first approximation does not depend on horizontal coordinates and if we replace density by the average value in the layer from $0 - H$, then from the equation for the field of mass in the midsection of a channel bounded at infinity it follows that

$$H = \sqrt{\frac{2\left(-\Omega \operatorname{curl}_z \vec{S} + S_x \frac{\partial \Omega}{\partial y} - \operatorname{div} \vec{T}\right)}{g\left(\frac{\partial^2 \bar{\rho}}{\partial x^2} + \frac{\partial^2 \bar{\rho}}{\partial y^2}\right)}}. \qquad (1.36)$$

In this expression H is the depth of the lower boundary of the baroclinic layer; $\Omega = 2\omega \sin \varphi$ is the coriolis parameter; $\vec{S} = \int_0^H \vec{u} dz$ is the full current characterizing the integral transport of water within the baroclinic layer; S_x is its projection on the coordinate axis X; T is the tangential wind stress; and $\bar{\rho}$ is the average sea water density along the vertical

$$\operatorname{curl}_z \vec{S} = \frac{\partial S_y}{\partial x} - \frac{\partial S_x}{\partial y};$$
$$\operatorname{div} \vec{T} = \frac{\partial T_x}{\partial x} + \frac{\partial T_y}{\partial y}.$$

The origin of the coordinates is at the sea surface, the Z axis is directed vertically downward, the X axis is directed to the east, and the Y axis, to the north.

From formula (1.36), obtained by V. B. Shtokman, it follows that the depth of the lower boundary of the baroclinic layer in the sea depends on the curl of the current in the density field, on the coriolis parameter and its latitudinal variation, and on divergence in the tangential wind stress field and the structure of the water density field. As known, the magnitude of a full current is indirectly associated with the curl of the tangential wind stress field and with the coefficient of lateral turbulent friction μ_x (Shtokman, 1949). If we make the proper substitution, then the denominator of formula (1.36) will have the coefficient of lateral turbulent friction μ_x. The value of μ_x depends on the vertical water density gradient. The greater is the stratification of sea water, the more intense is the horizontal turbulent exchange of the amount of motion.

If we consider that the "dominant" term in respect to magnitude among the terms in the numerator of formula (1.36) is $\mathrm{curl}_z S$, and the other two terms are small, we find that the depth of the lower boundary of the baroclinic layer depends on the curl in the tangential wind stress field and on water stratification. It increases with increasing $\mathrm{curl}_z T$ and with a decreasing vertical density gradient. Moreover, H depends on the geographical latitude of the area. Under otherwise equal conditions it is least at the equator and increases toward the poles.

These qualitative results clearly explain why the gradient current attenuates more rapidly in a strongly stratified sea than it does in a more homogeneous sea. The variation of H with latitude agrees with the conclusions reached by Defant (1941a) after analysis of a great number of observations in the Atlantic Ocean as well as with the conclusion of G. Neumann (1955). But V. B. Shtokman's formula gives only a qualitative estimate of the depth of penetration of a wind-driven current and is unsuitable for practical computation.

In the second section, we examined some results of Lineikin's study of wind circulation in an infinite channel. From the solution it follows that both components of the velocity of a wind-driven current decrease exponentially with depth and become zero only in infinity. The pure drift component of current velocity rapidly becomes insignificant whereas the gradient component

4. BAROCLINIC LAYER OF THE SEA

attenuates rather slowly with depth. Thus, the depth of penetration of a wind-driven current is limited by the gradient component of current velocity. To find the lower boundary of the baroclinic layer, we must investigate the pattern of the vertical variation of the gradient component of current velocity, while we need only to examine the gradient component to determine the "zero" surface.

The fact that the velocity of a wind-driven current becomes zero only at an infinite depth makes the concept of the lower boundary of the baroclinic layer, or the depth of penetration of the current, uncertain. Strictly speaking, H is always infinitely large. But as we can see from expressions (1.22) and (1.23), current velocity becomes small at great depths. From a practical standpoint, we are interested in current velocity to a definite limit, let us say to 1-2 cm/sec. Therefore, the isopleth with a given low current velocity v_0 can be considered as the depth of penetration of the current (the lower boundary of the baroclinic layer).

Then from (1.22) and (1.23) we obtain the following formula for determining the depth of the $v_0 = \text{const}$ surface:

$$H_{v_0} = \frac{L}{\pi \sqrt{bg}} \sqrt{4\omega_z^2 + \frac{\pi^4 \mu_x^2}{\rho^2 L^4}} \ln \frac{LT_0 \sqrt{bg} \cos(\varphi - 2\psi) \sin \frac{\pi x}{L}}{\pi v_0 \mu_x \sqrt{4\omega_z^2 + \frac{\pi^4 \mu_x^2}{\rho^2 L^4}}} . \quad (1.37)$$

All the designations are the same here as in section 2.

Formula (1.37) is considerably simplified if we consider that with large L at temperate and high latitudes

$$\frac{\pi^4 \mu_x^2}{\rho^2 L^4} \ll 4\omega_z^2. \quad (1.38)$$

Relationship (1.38) is invalid only in a narrow region near the equator. Beginning from a latitude of 5-6° inequality (1.38) is valid for all actual channel dimensions. Instead of (1.37) we then obtain

$$H_{v_0} \approx \frac{2\rho \omega_z}{\pi \sqrt{bg}} \ln \frac{LT_0 \sqrt{bg} \cos(\varphi - 2\psi) \sin \frac{\pi x}{L}}{2\pi \omega_z \mu_x v_0} . \quad (1.39)$$

The depth of the surface at which $v_0 = $ const is directly dependent on channel width L, on the tangential wind stress T_0, and on the vertical component of the angular speed of rotation of the earth ω_z (latitude of the area). It is inversely dependent on the coefficient of lateral turbulent friction μ_x and on the average vertical water density gradient b.

Following Ekman, who called "depth of frictional resistance" that depth at which the pure drift component of current velocity is e^π times smaller than at the sea surface, P. S. Lineikin introduced the concept of the "depth of baroclinicity," i.e., that depth at which current velocity is the $e^{-\pi}$ part of the gradient component of current velocity at the sea surface

$$\frac{v_{G\,(z=0)}}{v_{G\,(z=H)}} = e^{-\pi}. \tag{1.40}$$

In this case the "depth of baroclinicity" H is also determined by the relationship

$$H = L\sqrt{\frac{4\omega_z^2}{bg} + \frac{\pi^4 \mu_x^2}{bg\rho^2 L^4}}. \tag{1.41}$$

The second term in expression (1.41) can be neglected beyond the near-equatorial region without introducing any noticeable error. Then

$$H = \frac{2\omega L \sin \varphi}{\sqrt{g}} \cdot \frac{1}{\sqrt{b}}. \tag{1.42}$$

Here ω is the angular speed of rotation of the earth and φ is the geographical latitude of the area.

The values for the $\frac{2\omega L \sin \varphi}{\sqrt{g}}$ factor calculated from formula (1.42) with some given L and φ are presented in Table 2.

To obtain H we must divide the values given in the table by $\sqrt{b}$. Ordinarily b is of the order of 10^{-8}. That means that the "depth of baroclinicity" in meters is greater than the values given in the table by approximately a factor of 100.

4. BAROCLINIC LAYER OF THE SEA

Table 2

L, km	\multicolumn{9}{c}{φ, degree}
	10	20	30	40	50	60	70	80	90
1	0.08	0.16	0.23	0.30	0.36	0.40	0.44	0.46	0.47
10	0.80	1.60	2.34	3.00	3.58	4.05	4.39	4.60	4.68
100	8.00	16.0	23.4	30.0	35.8	40.5	43.9	46.0	46.8
1000	80.0	160	234	300	358	405	439	460	468
10,000	800	1600	2340	3000	3580	4050	4390	4600	4680

Lineikin's formulas establish the relationship between the depth of the lower boundary of the baroclinic layer and channel width, the geographical latitude of the area, the vertical water density gradient, and wind velocity. But they cannot be used to estimate even the depth of penetration of a current in a real sea. The averaged vertical water density gradient b in the formulas depends on H. This difficulty can be avoided by successive approximations in estimating H and b. But the fact that the depth of the isoline v_0 and the depth of baroclinicity are directly proportional to the width of the channel makes computation under actual conditions impossible. It becomes difficult to select L. If we estimate the depth of penetration of a current into the ocean, L evidently cannot be equal to the width of the ocean since in that case H would be too large. It is most reasonable to equate the size of the channel to the width of the wind zone. But as P. S. Lineikin shows for a narrow wind belt over the sea bounded by one shore, a wind-driven current extends beyond the zone under the influence of wind and maximum H is outside the wind belt.

To what extent the "depth of baroclinicity" depends on channel width L can be seen from the following example. At $40°$ latitude H is of the order of 3000 m in a channel 100 km wide; it is 30 km with $L = 1000$ km and 300 km with $L = 10,000$ km. This example shows clearly that theoretical conclusions based on conditions in an infinite channel cannot be applied to an actual sea without the danger of fundamental error. The best agreement between theory and actual conditions can be expected in a narrow strait that connects two large sea basins. In this case, actual conditions would satisfy the requirements of an infinite channel rather well.

In spite of the foregoing reservations, Lineikin's study is of great value, since it verifies theoretically a number of propositions that formerly were considered as assumptions, guesses, or hypotheses. For our purpose, the following fact is most important. The pure drift component of the velocity of a wind-driven current at sea is sufficiently well described in a stationary case by formulas of Ekman's classical theory. Consideration of turbulent friction does not lead to significant changes in the pure drift component of current velocity. The direction of the gradient component of current velocity does not vary along the vertical but the velocity of the gradient component attenuates with depth, reaching zero only at infinity. The influence of friction on the gradient current is manifest only as an insignificant deviation of the velocity vector from the direction of the geostrophic current. In the presence of a steady wind-driven current alone, the slopes of isobaric and isopycnal surfaces are opposite in sign. Aside from this, the Helland-Hansen and Ekman theorem (1923) of parallel solenoids in the sea must be satisfied in a stationary wind-driven current because the gradient component of current velocity has the same disturbing effect on the field of any physical-chemical property of sea water. In the velocity field of a steady wind-driven current, temperature, salinity, and density isolines, etc., must represent mutually parallel lines in any section.

Strictly speaking, there is no lower boundary to the baroclinic layer in the velocity field of a wind-driven current. We can only speak of a conventional lower boundary of a current, i.e., of the depth of baroclinicity or of the depth of the v_0 isoline. It is quite evident that when the gradient component of the velocity of a steady wind-driven current is computed by the dynamic method, the current velocity obtained will be closer to the actual velocity the deeper the reference surface becomes, since $\vec{v} \to 0$ with $H \to \infty$.

But if, in addition to a wind-driven current, a convective current of advective origin is induced by the transport of heat and salt, their combined effect could produce in a number of cases a steep slope in the boundary

4. BAROCLINIC LAYER OF THE SEA

between the two currents and thus lead to the formation of a lower baroclinic layer boundary at a depth $H \neq \infty$. Let us demonstrate this by an example. Assume that wind action at the sea surface produces two adjacent currents of opposite direction. The diagram in Fig. 7a shows the vertical cross section through part of the sea. Solid lines represent the traces of isobaric surfaces with an exaggerated slope for clarity. On the left of the diagram the gradient current is directed from the drawing and on the right, into the drawing. The dashed lines represent traces of isopycnal surfaces. The gradient component of current velocity is highest at the surface of the sea and approaches zero with increasing depth.

Again, let us examine a convective current induced by mass transport in a wind-driven current. Without considering the wind-driven current, let us assume that the right side of the vertical cross section is filled with lighter water and the left side with heavier water. In this case isopycnal surfaces must slope from left to right (diagram b in Fig. 7). Since the transverse difference in density is not associated with heating or cooling or precipitation and evaporation in the given region, but results from water transport, we can assume that this horizontal inhomogeneity in water density occurs without producing change in the relief of the free sea surface. Consequently, the sea surface in a convective current zone is assumed to be horizontal. The slopes of isobaric surfaces increase with depth, approaching a limiting value. The slopes of isobaric surfaces created by the horizontal inhomogeneity of sea water density are evidently the same as the slope of isopycnal surfaces. The convective current is directed into the diagram in conformity with the structure of the water density field represented in Fig. 7b. Its velocity is zero at the sea surface and increases with depth to some limiting value.

Let us recall again that the convective component of current velocity under consideration is not independent and its distinction from the total gradient current is artificial.

Mass transport in a wind-driven current merely disturbs the pattern of vertical variation of the current's velocity. In effect, this amounts to the presence of an additional current velocity component.

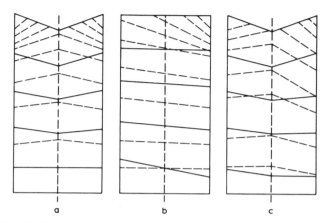

Fig. 7. Traces of isobaric (solid lines) and isopycnal (dashed lines) surfaces:
a—in the wind-driven current; b—in a convective current of advective origin; c—their total effect.

Let us now examine the combined effect of the gradient component of the velocity of a wind-driven current and a convective current (Fig. 7c). To do this, we must find what results can be obtained by adding these two components of current velocity, proceeding from general considerations. In the left part of the diagrams in Fig. 7a and b the slopes of isopycnal surfaces are opposite in sign. Consequently, mass advection interferes with the normal process of the readjustment of the density field in the cold current region and tends to weaken the vertical change in the velocity of the combined current in this part of the cross section. Similar slopes of isobaric surfaces in the left part of the cross section in the first two diagrams confirm the above.

This means, then, that adding the gradient component of the velocity of a wind-driven current with a convective current in a denser water mass must lead to an increase in current velocity in the layers beneath the surface, to a reduction of the intensity of its vertical variation, and to a decrease of the slopes of isopycnal surfaces.

In the right part of the cross section, which contains less heavy water, the isopycnal surfaces slope to one side

4. BAROCLINIC LAYER OF THE SEA

in the first two diagrams of Fig. 7. Consequently, mass transport produces conditions that promote an increase in the intensity of vertical current velocity variation in the region of less heavy water. It can be expected that the gradient component of the velocity of the wind-driven current and the convective current would fully compensate each other at some depth. If below this horizon there remains a transverse inhomogeneity in water density, then a countercurrent will form in the underlying water mass.

Current velocity curves are shown in Fig. 8. Fig. 8c gives curves of the velocity of a wind-induced gradient current in the region of denser water and Fig. 8a, in the region of less dense water. Vertical stratification is greater in less dense than in more dense water. Therefore, current velocity attenuates more slowly with depth in the left part of the cross section than in the right. Figure 8b shows velocity curves of the convective current induced by the horizontal inhomogeneity of the water density field as a result of mass advection in the wind-driven current. Data in Table 1 were used to construct the last curve.

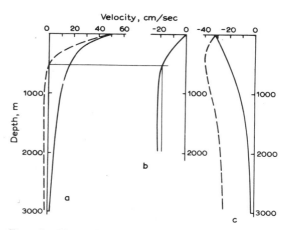

Fig. 8. Vertical distribution of gradient current velocity:
a—wind-driven current in the warm water mass; b—convective current of advective origin; c—wind-driven current in the cold water mass; the dashed lines represent curves of the total current.

Total velocity in a current carrying denser water is found by simple addition of the convective and gradient components of current velocity since they have the same direction. As we can see from the diagram, the tendency of the velocity of the total current to attenuate with depth becomes very small in the region of heavier water. Therefore, we can expect the current to reach bottom.

In the region of less dense water in the right part of our diagrams, the gradient component of the velocity of the wind-driven current and the convective current have opposite directions. Therefore, simple addition of current velocity vectors can be performed only to horizon H at which they mutually compensate each other. This conclusion follows from the fact that both current velocity components under consideration are gradient components by nature. The density inhomogeneity in the water mass overlying $z > H$ horizons does not play a part in the formation of a horizontal pressure gradient at these horizons because the effect of the slope of the sea surface is fully compensated by the horizontal inhomogeneity of the density field at depth H and the horizontal pressure gradient is thus zero. Therefore, current at the $z > H$ horizons is induced only by the inhomogeneity of the density field. To obtain current velocity at these deeper horizons, the curve for the convective component of current velocity should be read from point $z > H$. This is shown in Fig. 8b by additional traces.

Having examined a particular idealized example, we have come to a very interesting conclusion. Current velocity attenuates slowly with depth and the current may reach bottom in the denser water mass with almost horizontal or even horizontal isopycnal surfaces in the region where two currents carrying water of different densities meet (discontinuity or front). The slopes of isopycnal surfaces are steep and univalent in the entire layer of less dense waters; current velocity changes rapidly with depth; and the current may reverse starting at some subsurface horizon.

These conclusions are in contradiction to the existing view that gently sloping isopycnal surfaces testify to the low velocity of the gradient current. The foregoing example shows that in some cases the slope of isopycnal surfaces testifies to the contrary, but is always directly associated with the intensity of the vertical variation of the gradient component of current velocity. The larger is the horizontal water density gradient the greater the vertical change of the gradient component of the velocity of a steady current. Good examples of the three manifestations of wind action on the sea surface are found in those regions where the Gulf Stream and the Labrador current meet in the Atlantic and the Kuroshio and Oyashio meet in the Pacific (discontinuities, fronts, or "cold walls"). The wind action manifestations are: pure drift resulting directly from the entraining action of the wind; gradient current produced by the slope of the free surface of the sea and the adjustment of the water density field; and convective current resulting from mass transport in a wind-driven current. Figure 9 shows the distribution of density expressed as σ_t along the profile from the Chesapeake Bay to the Bermuda Islands according to data collected by the *Atlantis*. The core of the Gulf Stream is located in the region of station 1228, i.e., in the region where isopycnal surfaces are steepest. Figure 9 and the third diagram in Fig. 7 agree surprisingly well. In the denser water mass the isopycnal surfaces in both diagrams are almost horizontal, while in the region with less heavy water the isopycnal surfaces are steep. If we use the hypothetical example above to explain this fact, we must conclude that in the left part of the section, in the denser water mass, current velocity changes little along the vertical, but can be very high. Current velocity can be high even in the bottom layer. Then it becomes clear that the current computed by the dynamic method will be too low. Current velocity in the Gulf Stream changes more intensely with depth and reaches low values much more rapidly. There is reason to expect the development of a countercurrent in the deep layers. The decrease in the slope of isopycnal surfaces to the right of the Gulf Stream also can be explained from the foregoing considerations, since the water there is relatively denser than the water of the Gulf Stream itself.

It can be shown that in those regions where the transport of heat and salt by a wind-driven current results in an additional convective component of current velocity, the conditions of the Helland-Hansen and Ekman theorem

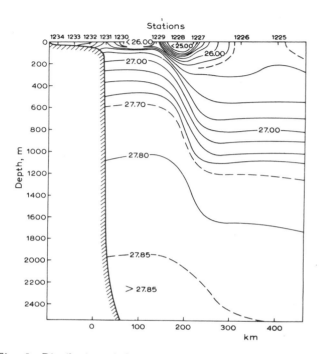

Fig. 9. Distribution of density σ_t along the Atlantis profile from the Chesapeake Bay to Bermuda.

cannot be satisfied. In this case, the temperature, salinity, and density isolines will not coincide in a vertical or any other section. This becomes obvious if we consider the well-known observation that, as a rule, a warmer water mass has a higher salinity than a colder one. An identical change in water temperature and salinity has an inverse effect on water density.

If in some ocean region there is only a stationary wind-driven current, the theorem concerning the parallelism of solenoids beyond the upper frictional layer must be satisfied accurately enough, since the current velocity field has the same disturbing influence on the distribution of any physical-chemical property of sea water. In this case, the lower boundary of the baroclinic layer is some conventional surface and the reference surface must be placed as deep as possible when computing current velocity by the dynamic method. This is because current velocity tends toward zero with $z \to \infty$.

If the conditions of the Helland-Hansen and Ekman theorem are satisfied, it becomes difficult to compute the gradient component of the velocity of a steady sea current by the dynamic method. In fact, the nonparallelism of the isolines of water properties indicates that in the region under study the structure of the water density and current velocity fields can be widely distorted because of mass transport by the wind-driven current. This could lead, first, to the development of a countercurrent in the deep layers and, second, to excessive or low current velocity as computed by the dynamic method.

CONCLUSIONS

1. The gradient component of a steady sea current is computed by the dynamic method from the distribution of sea water density. This component of current velocity is induced by the horizontal pressure gradient which depends on the slope of the free surface of the sea and on the horizontal inhomogeneity of the water density field.

2. Wind action on the sea surface and the nonuniformity of climatic conditions over the sea produce a slope in the sea surface, lead to the horizontal inhomogeneity of water density and to the appearance of a gradient current. The total steady gradient current is computed by the dynamic method.

3. The dynamic relief of the sea surface or any other isobaric surface is the same as its geometric relief expressed on a different scale. To convert dynamic relief to geometric relief it is sufficient to divide the dynamic heights of the isobaric surface by the acceleration of gravity.

4. The "zero" surface is located at that depth in the sea where the two components of the horizontal pressure gradient, one due to the slope of the free surface of the sea and the other, to the horizontal inhomogeneity of the mass field, are mutually compensated.

5. In a baroclinic sea, the horizontal pressure gradient created by the slope of the sea surface and the inhomogeneity of the water density field in the layer overlying the horizon under consideration is equal in absolute value and opposite in sign to the horizontal pressure gradient produced by the inhomogeneity of water density in the underlying layer (from the given horizon to the "zero" surface) in the steady state.

6. If the conditions of the theorem of the parallelism of solenoids in the sea are satisfied, the gradient current penetrates to an infinitely great depth and there is no strict "zero" surface. Current velocities computed by the dynamic method agree better with absolute values the deeper the reference surface is placed.

7. The depth to which a current penetrates in the sea depends on the geographical and synoptic conditions of the entire basin. It increases with increasing wind velocity and curl, with increasing geographic latitude, and with decreasing water stratification.

CHAPTER II

Accuracy of the Dynamic Method of Computing Ocean Currents

5. STATEMENT OF THE PROBLEM

After examining the results of theoretical investigations of ocean currents in the preceding chapter, we concluded that the gradient current is nearly geostrophic when the law of the parallelism of solenoids in the sea is satisfied. The velocity of this current decreases with depth, and approaches zero asymptotically. Consequently, in this case dynamic computation requires that the reference surface be placed as deep as possible; also, both probability and numerous direct measurements show that, as a rule, current velocity decreases with increasing depth.

In some cases, however, the reference surface may be too deep; the dynamic method yields a picture of horizontal water circulation that contradicts the concept of currents in the region under study. This occurs most frequently in seas with a weak horizontal circulation. For this reason it is interesting to find the limits of accuracy of the dynamic method of computation and to determine the influence of individual factors on its reliability. As far as we know, disregarding a few general comments on this problem (Thompson, 1939; Jakhelln, 1936; Dobrovol'skii, 1949), nobody has studied the accuracy of the dynamic method in detail.

The accuracy of the dynamic method as a function of the accuracy of physical measurement will be examined below. All arguments and conclusions will be based on the accuracy of measurement of initial values, as determined by errors in the scale reading of measuring equipment; we shall not consider systematic and subjective errors.

No theoretical distinction is usually made between "error" and "mistake." The expressions "measurement error" and "mistake in measurement," and "computation error" and "mistake in computation" are considered identical. For the sake of convenience, let us designate as "mistake" the maximum possible error in measurement or computation, thus avoiding the necessity of indicating each time what kind of error is meant.

Before discussing the accuracy of the dynamic method, we must determine the limits of allowable errors in the determination of initial values: the accuracy of temperature and salinity measurements and the accuracy of sea water density computations.

6. ACCURACY OF SEA WATER TEMPERATURE MEASUREMENTS

The temperature of sea water at various horizons is usually measured with deep-water reversing thermometers. Modern deep-water thermometers have minimum scale divisions of $0.1°C$. The reading accuracy of such a scale is $\pm 0.02°C$. Temperature is read with an accuracy of $\pm 0.1°C$ from an auxiliary thermometer. A reduction correction is then introduced into the thermometer readings. Formulas for computing the reduction correction for temperature and its relative error are:

$$T = T' + k, \tag{2.1}$$

$$k = \frac{(t - T')(T' + v_0)}{n}\left(1 + \frac{T' + v_0}{n}\right), \tag{2.2}$$

$$\frac{dk}{k} = \frac{dt}{t - T'} + \left(\frac{1}{T' + v_0} - \frac{1}{t - T'} + \frac{1}{n + T' + v_0}\right)dT' + \\ + \left(\frac{1}{T' + v_0} + \frac{1}{n + T' + v_0}\right)dv_0 + \left(\frac{T' + v_0}{n + T' + v_0} + 1\right)\frac{dn}{n}. \tag{2.3}$$

In these expressions T is the corrected water temperature; T' is the reading of the main thermometer; k is the reduction correction for temperature; t is the reading of the auxiliary thermometer; n is the coefficient of the relative volumetric expansion of mercury and of the thermometer glass; v_0 is the volume of mercury in the upper wide part of the thermometer up to the 0° graduation mark, as expressed in degrees of the scale of the main thermometer.

From expression (2.3) it follows that the relative error in the determination of the reduction correction for temperature depends on the accuracy of the readings of the main and auxiliary thermometers, as well as on the allowable errors in the determination of the volume of mercury v_0 and of the coefficient of volumetric expansion of mercury and of the thermometer glass n. To estimate the error introduced by the reduction correction of the sea water temperature, let us assume that $dv_0 = \pm 1°C$ and $\frac{dn}{n} = \pm 0.01$. The volume of mercury v_0 and the coefficient of volumetric expansion n are determined when the thermometer is manufactured. Their accuracy is apparently quite reliable.

Computation by the above formulas shows that in most cases the reduction correction does not introduce additional errors into the sea water temperature. The reduction correction may lower the accuracy of temperature measurements by $\pm 0.01°C$ only if $t - T'$ differences are very large and v_0 is large. In this case, as we can see from formula (2.3), the relative error in the computation of the reduction correction decreases but the reduction correction increases, since the absolute error of dk increases.

Thus, the temperature of sea water is measured by deepwater reversing thermometers with an accuracy equal to the accuracy of the scale reading on the main thermometer, $dT = \pm 0.02°C$. On rare occasions the error in the determination of water temperature may reach $\pm 0.03°C$.

7. ACCURACY OF SEA WATER SALINITY (CHLORINITY) DETERMINATIONS

The method most widely used in oceanology for determining the salinity of sea water from chlorine is the chemical method. The chlorine and salt content of sea water are computed from the formulas below, which were used to compile special tables (Zubov, 1957).

$$S = 0.030 + 1.8050\, Cl; \tag{2.4}$$

$$dS = 1.8050\, dCl; \tag{2.5}$$

$$Cl = a + k'; \tag{2.6}$$

$$dCl = da + dk'; \tag{2.7}$$

$$k' = \frac{a\left(R - \frac{\rho_{17.5}}{1000}\right)(1-R)}{1 - \left(R - \frac{\rho_{17.5}}{1000}\right)(1-R)}; \tag{2.8}$$

$$\frac{dk'}{k'} = \frac{da}{a} + \left[\frac{1000}{1000\,R - \rho_{17.5}} - \frac{1}{1-R} + \frac{1 - 2R + \frac{\rho_{17.5}}{1000}}{1 - \left(R - \frac{\rho_{17.5}}{1000}\right)(1-R)}\right]dR +$$
$$\left\{\frac{1}{1000\,R - \rho_{17.5}} + \frac{1-R}{1000\left[1 - \left(R - \frac{\rho_{17.5}}{1000}\right)(1-R)\right]}\right\}d\rho_{17.5}; \tag{2.9}$$

$$R = \frac{\rho'_{17.5}}{1000} + \left(\frac{N}{A} + 1\right)\left(1 + \frac{\rho'_{17.5}}{1000}\right); \tag{2.10}$$

$$dR = \frac{N}{1000\,A}\,d\rho'_{17.5} + \frac{1000 + \rho'_{17.5}}{1000\,A}\,dN + \frac{N(1000 + \rho'_{17.5})}{1000\,A^2}\,dA. \tag{2.11}$$

In these expressions S is the per-mille salinity of sea water; Cl is the chlorine content per mille; a is the buret reading in titration of sea water; $\rho_{17.5}$ is the specific gravity of normal water at a temperature of 17.5°C and atmospheric pressure; $\rho'_{17.5}$ is the specific gravity of the given sea water at a temperature of 17.5°C and atmospheric pressure; N is the chlorine content in normal water per mille; and A is the reading of the buret during titration of normal water.

We now examine the formulas for mistakes (the limiting magnitude of errors) in order to determine the accuracy

7. SEA WATER SALINITY DETERMINATIONS

of sea water chlorinity and salinity determinations. Expressions (2.10) and (2.11) include values that change very little when applied to the ocean. Therefore it is possible to estimate the error of dR by substituting numerical values in formula (2.11). Neglecting relatively small terms, we obtain

$$dR = 0.05 \cdot dA. \qquad (2.12)$$

Similarly, the relative error in the determination of k' is

$$\frac{dk'}{k'} = 139 \cdot dR. \qquad (2.13)$$

By combining the last two expressions and substituting the result in (2.7), we obtain an expression for estimating the error in the determination of sea water chlorinity from titration data

$$dCl = da + dA. \qquad (2.14)$$

The accuracy of determinations of the chlorine content in water depends on the accuracy of buret readings in titration of samples of a given sea water da and normal water dA; the coefficients are very close to one for the values in formula (2.14). The allowable error in titration of normal water can be neglected for our purpose. We can easily demonstrate that it is significant only when salinity at two neighboring hydrological stations is determined with different silver nitrate solutions, or, more accurately, with different solution titers.

The smallest scale division on a buret is usually 0.01 ml. If the accuracy of thermometer readings is assumed to be one-fifth of a scale division, the buret can be read only with an accuracy of one-half of a scale division, because the meniscus interferes with the reading. Then $da = \pm 0.005$ ml, the chlorine content in sea water (2.14) is determined with an error of $\pm 0.005°/_{00}$, and the error in the computation of salinity from chlorine is $\pm 0.009°/_{00}$. It should be remembered that the errors chosen are valid only for estimating the accuracy of the dynamic method. The accuracy of absolute chlorinity and salinity determinations must be half that, even if the empirical relationships used as the basis of computation are valid.

8. ACCURACY OF SEA WATER DENSITY COMPUTATIONS

Potential sea water density, σ_t, depends on water temperature and salinity (chlorinity). This relationship is expressed by an empirical formula (Bjerknes and Sandström, 1912). Since this formula is cumbersome, let us write only the expression obtained from this formula needed for estimating the error in the computation of σ_t from given errors in the determination of sea water temperature and chlorinity

$$d\sigma_t = \alpha\, dCl + \beta\, dT, \qquad (2.15)$$

where

$$\alpha = (1.4708 - 0.003140\, Cl + 0.0001194\, Cl^2)(1 - A_t + 2\sigma_0 B_t); \qquad (2.16)$$

$$\beta = (\sigma_0 + 0.1324)\left[(\sigma_0 - 0.1324)(18.030 - 1.6328\, T + 0.05001\, T^2) \cdot 10^{-6} - (4.7867 - 0.196370\, T + 0.0032529\, T^2) \cdot 10^{-3} - \sum_t \left(\frac{2}{T - 3.98} + \frac{1}{T + 283} - \frac{1}{T + 67.26}\right)\right]. \qquad (2.17)$$

In the last two expressions

$$A_t = (4.7867\, T - 0.098185\, T^2 + 0.0010843\, T^3) \cdot 10^{-3}; \qquad (2.18)$$

$$B_t = (18.030\, T - 0.8164\, T^2 + 0.01667\, T^3) \cdot 10^{-6}; \qquad (2.19)$$

$$\sigma_0 = -0.069 - 1.4708\, Cl - 0.001570\, Cl^2 + 0.0000398\, Cl^3; \qquad (2.20)$$

$$\sum_t = \frac{(T - 3.98)^2}{503.570} \cdot \frac{T + 283}{T + 67.26}. \qquad (2.21)$$

Coefficients α and β in formula (2.15) vary, depending on sea water temperature and chlorinity. Consequently,

8. SEA WATER DENSITY COMPUTATIONS

the error in computing $d\sigma_t$ also varies. Coefficients α and β at various water temperatures and chlorinities are presented in Table 3.

Table 3

Coefficient α

Cl, ‰	\multicolumn{9}{c}{T, degrees}

Cl, ‰	−2	0	2	5	10	15	20	25	30
5	1.5	1.5	1.5	1.4	1.4	1.4	1.4	1.4	1.3
6	1.5	1.5	1.4	1.4	1.4	1.4	1.4	1.4	1.3
7	1.5	1.5	1.4	1.4	1.4	1.4	1.4	1.4	1.3
8	1.5	1.5	1.4	1.4	1.4	1.4	1.4	1.3	1.3
9	1.5	1.5	1.4	1.4	1.4	1.4	1.4	1.3	1.3
10	1.5	1.5	1.4	1.4	1.4	1.4	1.4	1.3	1.3
11	1.5	1.5	1.4	1.4	1.4	1.4	1.4	1.3	1.3
12	1.5	1.5	1.4	1.4	1.4	1.4	1.4	1.3	1.3
13	1.5	1.5	1.4	1.4	1.4	1.4	1.4	1.4	1.3
14	1.5	1.5	1.4	1.4	1.4	1.4	1.4	1.4	1.3
15	1.5	1.5	1.4	1.4	1.4	1.4	1.4	1.4	1.3
16	1.5	1.5	1.4	1.4	1.4	1.4	1.4	1.4	1.4
17	1.5	1.5	1.4	1.4	1.4	1.4	1.4	1.4	1.4
18	1.5	1.5	1.4	1.4	1.4	1.4	1.4	1.4	1.4
19	1.5	1.5	1.4	1.4	1.4	1.4	1.4	1.4	1.4
20	1.5	1.5	1.4	1.4	1.4	1.4	1.4	1.4	1.4

Coefficient β

Cl, ‰	−2	0	2	5	10	15	20	25	30
5	0.7	0.4	0.2	−0.1	−0.6	−1.1	−1.6	−1.8	−2.2
6	0.9	0.5	0.2	−0.2	−0.8	−1.4	−1.9	−2.3	−2.2
7	1.0	0.6	0.3	−0.2	−0.9	−1.6	−2.3	−2.6	−3.1
8	1.2	0.7	0.3	−0.2	−1.1	−1.8	−2.6	−3.0	−3.6
9	1.3	0.8	0.4	−0.3	−1.2	−2.0	−2.9	−3.4	−4.0
10	1.5	0.9	0.4	−0.3	−1.3	−2.2	−3.2	−3.8	−4.5
11	1.6	1.0	0.5	−0.3	−1.5	−2.5	−3.5	−4.1	−4.9
12	1.8	1.1	0.5	−0.4	−1.6	−2.7	−3.9	−4.5	−5.4
13	1.9	1.2	0.6	−0.4	−1.7	−2.9	−4.2	−4.9	−5.8
14	2.1	1.3	0.6	−0.4	−1.9	−3.1	−4.5	−5.3	−6.2
15	2.2	1.4	0.6	−0.4	−2.0	−3.4	−4.8	−5.6	−6.7
16	2.4	1.4	0.7	−0.5	−2.1	−3.6	−5.2	−6.0	−7.0
17	2.5	1.5	0.7	−0.5	−2.3	−3.8	−5.5	−6.4	−7.6
18	2.6	1.6	0.8	−0.5	−2.4	−4.0	−5.8	−6.8	−8.0
19	2.8	1.7	0.8	−0.6	−2.6	−4.2	−6.1	−7.1	−8.5
20	2.9	1.8	0.8	−0.6	−3.2	−4.5	−6.4	−7.5	−8.9

The sign of coefficient β does not have to be considered since the error in temperature measurements can be either positive or negative.

The table shows that the error in the computation of sea water density depends on salinity and varies considerably with temperature. Thus, with $Cl = 20\,^o/_{oo}$ the error in potential density determination is ± 0.02, ± 0.09, and ± 0.19 in σ_t units with $T = 2°$, $15°$, and $30°C$, respectively.

As a rule, high water temperatures are observed only in a relatively thin surface layer in temperate latitudes. Below this layer water temperature approaches zero.* Salinity, on the other hand, generally varies within wide limits. Hence we shall assume that the ultimate error in the determination of sea water density is constant and equal to ± 0.02 σ_t units. This assumption is obviously unrealistic, but since we intend to give a qualitative rather than a quantitative estimate of the accuracy of the dynamic method, it simplifies further analysis considerably. For other purposes, or for a more detailed examination of the accuracy of the method, the error in the determination of density can be easily computed on the same basis as a variable.

The specific volume v_t, is more frequently used in dynamic computations. The computation error

$$dv_t = \frac{d\sigma_t \cdot 10^6}{(\sigma_t + 1000)^2} \qquad (2.22)$$

is approximately 95% of the error in the computation of potential density, i.e., it also averages $+0.02$.

Let us note that density corrections to allow for the compressibility of sea water are computed with an accuracy exceeding that of density computations by approximately two orders of magnitude. Consequently, introduction of these corrections does not change the computed error in the determination of sea water density.

---

*Most deep water in the world's oceans is warmer than 0°C. —Ed.

9. ACCURACY OF DYNAMIC HEIGHT COMPUTATIONS

The dynamic height of an isobaric surface p is found from the expression

$$D = \int_0^p v_t(p) \cdot dp, \tag{2.23}$$

where D is dynamic height and $v_t(p) \equiv v_t(z)$ represents the vertical distribution of specific volume. The origin of coordinates is located on the reference surface and the p axis is directed vertically upward.

When computing dynamic height, the pressure measured in decibars is usually assumed to be equal to depth z, which is measured in geometric meters. The vertical distribution of specific volume can be considered to be the same in both cases, since $p \approx 1.01\ z$. The error in dynamic height obtained as a result of substitution of integration variables is 1% of its value and, therefore, substitution of depth for pressure increases the computed current velocity by 1%. This error is relatively small and can be neglected.

Since the vertical distribution of specific volume is given from tabulated observations, the integral in formula (2.23) is computed approximately, usually by the trapezoidal method

$$\int_0^z v_t(z)\, dz = \sum_m \frac{\gamma_n - \gamma_{n+1}}{2}[v_t(\gamma_n) + v_t(\gamma_{n+1})] - \sum_m \frac{(\gamma_n - \gamma_{n+1})\, v_t''(\eta)}{12}. \tag{2.24}$$

Here γ_n and γ_{n+1} designate the depths of two neighboring measurement levels; m is the number of layers between the reference surface ($z = 0$) and the given isobar ($z = p$); and η is some intermediate depth ($\gamma_n > \eta > \gamma_{n+1}$).

The second term on the right side of (2.24) represents the remainder term of the trapezoid formula. It describes the error of approximate integration due to the substitution of the smoothly changing function $v_t(z)$ by a broken line. Under actual conditions, the vertical variation in density is such that approximate integration does not introduce a significant error into dynamic height. An estimate of the remainder term indicates that it can be large only when the vertical density variation is very irregular. The vertical structure of the density field in the sea is usually complex only in the layers just below the surface, but there observations are generally closely spaced. Therefore we shall ignore this term.

From formula (2.24) it follows that the ultimate error in the computation of conventional dynamic height dD is directly proportional to the mistake in the determination of specific volume dv_t and to the distance H from the reference surface to the given isobar

$$dD = H \cdot dv_t. \qquad (2.25)$$

H is expressed in meters here. As assumed earlier, the mistake in the determination of specific volume is constant and equal to ± 0.02 units.

Since in this case we are interested in the ultimate error in dynamic height, when deriving formula (2.25) we assumed that the mistake in specific volume dv_t is identical along the entire vertical. Moreover, we neglected the term that depends on the error in the determination of the depth of measurement. This is admissible because error-free precise depth increments Δz are used in computing dynamic heights. Obviously, an inaccuracy in the determination of the depth of the measurement level will affect only the reliability of the specific volume and the vertical distribution of the hydrological elements being measured. But since $dz \leqslant 0.02\ z$, we can safely assume that $v_t(z) = v_t(z \pm 0.02\ z)$ and neglect this error.

Let us now illustrate these results by an actual example. To do this, dynamic heights were calculated from direct measurements and a dynamic relief chart of the sea surface was constructed (Fig. 10a). The horizontal reference surface is located at a depth of 1000 m.

From formula (2.25) it follows that here the ultimate absolute error in the computation of dynamic heights is ± 20 dyn·mm, i.e., the true value of each dynamic height

9. ACCURACY OF DYNAMIC HEIGHT COMPUTATIONS

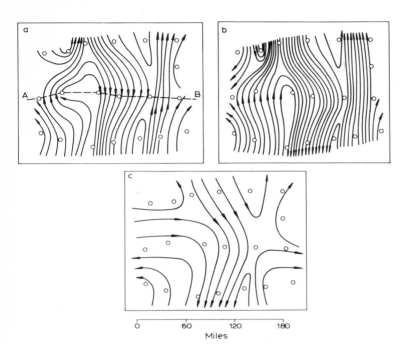

Fig. 10. Dynamic relief of the sea surface:
a—computed; b—exaggerated to the maximum; c—smoothed to the maximum.

differs from its computed value by not more than ± 20 dyn·mm.

Because the errors in the computation of dynamic heights differ (only their ultimate values are equal), the form of the actual dynamic relief can differ considerably from the computed relief and, consequently, the picture obtained by the dynamic method may differ significantly from actual horizontal circulation. It is evident that the differences will be greatest in regions with low current velocities, where the error in computation of dynamic height is comparable in magnitude to the characteristic value of the difference between dynamic heights at neighboring hydrological stations. There is reason to believe that in some cases computational errors may completely distort the picture of currents. The computational error cannot significantly influence the reliability of the computed current in regions where current velocities are high. Aside from regions with a clearly defined horizontal circulation, there are areas in the sea where current velocities are small. Therefore it is very important to estimate the reliability of each section of the dynamic chart when making dynamic computations. This is possible because of the accuracy of dynamic height computation.

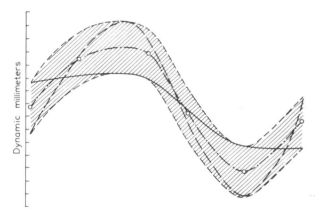

Fig. 11. Relief of the sea surface along section AB (Fig. 10).

A cross section of the dynamic relief of the surface of the sea along AB (Fig. 10a) is presented in Fig. 11. Each division of the vertical scale is equal to 10 dyn·mm. The interval of 20 dyn·mm from the computed dynamic profile (point-dash line) is hachured in the diagram. The curve of the actual dynamic profile of the sea surface must evidently be located within the hachured area, since the error in dynamic height at each hydrological station does not exceed 20 dyn·mm. The actual dynamic profile can be represented by any of the innumerable curves that can be drawn within this area. Let us examine two extreme cases: the most dissected and the smoothest dynamic relief. To find such profiles we merely draw the two curves in the hachured area corresponding to the two cases under

9. ACCURACY OF DYNAMIC HEIGHT COMPUTATIONS

consideration. The solid line in Fig. 11 represents the smoothest dynamic profile and the dashed line, the most dissected one. Comparison of the three dynamic profiles makes it possible to estimate the reliability of current velocity components normal to the cross section that are obtained by the dynamic method.

After similarly processing two other cross sections, we constructed dynamic charts of the surface of the sea for such cases when computation errors intensify (Fig. 10b) and smooth (Fig. 10c) the dynamic relief to the maximum. Even the most superficial comparison of the three dynamic charts (Fig. 10a, b, c) shows that the most reliable current is that which is directed from north to south in the center of the chart. The countercurrents in the western and eastern parts are less reliable, since they are hardly noticeable in the chart of the smooth dynamic relief.

It was assumed above that the error in the determination of specific volume is at its allowable maximum along the entire vertical. We examined the maximum possible error (i.e., the mistake) in dynamic height. Actually, we can hardly expect that in the computation of dynamic heights all errors summate and do not compensate each other, at least partially. We shall now demonstrate that the actual error in the computation of dynamic heights is approximately half the mistake.

We know from mathematical statistics that the probability of occurrence of error ϑ in the summation of s values, each of which may with equal probability contain an error equal to one of the numbers of the series

$$-l, -(l-1), -(l-2), \ldots, -2, -1, 0, 1, 2, \ldots \ldots, (l-2), (l-1), l, \quad (2.26)$$

is estimated by the expression

$$P(\vartheta) = \frac{h}{\sqrt{\pi}} \exp(-h^2\vartheta^2), \quad (2.27)$$

where

$$h = \frac{\sqrt{6}}{2\sqrt{l(l-1)s}}. \quad (2.28)$$

$P(\vartheta)$ is the probability of error ϑ and h is the statistical measure of accuracy.

Formulas (2.27) and (2.28) could be used to calculate the probability of errors of given magnitude in dynamic height or to calculate the magnitude of the error from a given probability if measurements of hydrological elements at sea were made at equal depth intervals Δz. Then the ultimate error of each of the s terms would be

$$l = \Delta z \cdot dv_t. \tag{2.29}$$

For the sake of simplicity, let us assume that temperature and salinity are measured at depth intervals of 25 m from the surface to 200 m, 100 m from 200-1000 m, and 500 m at greater depths. If the reference surface is located below 1000 m, the column of water from the sea surface to the reference surface is divided into three intervals. Formula (2.27) is applicable within each of these intervals because depth increments are constant within the interval.

We know from the theory of probability that the composition of two or more normal distributions is a normal distribution. Consequently, in this case the probability of error ϑ in the computation of dynamic height is

$$P(\vartheta) = \frac{Q}{\sqrt{\pi}} \exp(-Q^2 \vartheta^2), \tag{2.30}$$

where

$$Q = \frac{h_1 h_2}{\sqrt{h_1^2 + h_2^2}} \text{ for } H \leqslant 1000 \text{ m}, \tag{2.31}$$

$$Q = \frac{h_1 h_2 h_3}{\sqrt{h_1^2 h_2^2 + h_1^2 h_3^2 + h_2^2 h_3^2}} \text{ for } H \geqslant 1500 \text{ m}. \tag{2.32}$$

Here h_1, h_2, and h_3 are the statistical measures of accuracy in the first, second, and third depth intervals (0-200 m, 200-1000 m, and >1000 m), respectively. They are computed by formula (2.28).

From expressions (2.28), (2.30), (2.31), and (2.32) it follows that errors in dynamic height computation that are equal in absolute magnitude and opposite in sign appear

9. ACCURACY OF DYNAMIC HEIGHT COMPUTATIONS

with equal probability. The probability of an error decreases as its absolute magnitude increases. The probability of an infinitely large error is zero. The most probable error is equal to zero. All this testifies to the fact that we have obtained a normal random distribution of errors (Fig. 12).

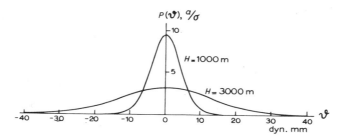

Fig. 12. Normal distribution of errors in the computation of dynamic heights.

As distance H from the reference surface to a given isobar increases, the statistical measure of accuracy, represented in Fig. 13 as a function of H, decreases. This leads to a decrease in the probability of small errors and to an increase in the probability of large errors in the computation of dynamic height.

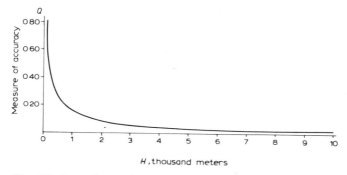

Fig. 13. Dependence of the statistical measure of accuracy on the depth of the reference surface.

Table 4

Probability of errors in the computation of the dynamic height of the sea surface, %

H, m

[ϑ] dyn·mm	500	600	700	800	900	1000	1500	2000	2500	3000	3500	4000	4500	5000
0	19.5	12.7	11.3	10.7	9.9	9.4	6.4	4.2	3.5	3.1	2.8	2.6	2.4	2.2
5	3.9	3.6	4.1	4.4	4.6	4.7	4.7	3.6	3.2	2.9	2.7	2.5	2.3	2.1
10		0.1	0.2	0.3	0.5	0.6	1.8	2.4	2.4	2.3	2.2	2.1	2.0	1.9
15							0.4	1.2	1.5	1.6	1.6	1.6	1.6	1.6
20								0.5	0.7	0.9	1.0	1.1	1.2	1.2
25								0.1	0.3	0.5	0.6	0.7	0.8	0.8
30									0.1	0.2	0.3	0.4	0.5	0.5
35										0.1	0.2	0.2	0.2	0.3
40											0.1	0.1	0.1	0.2
45														0.2
50														0.1
[dD], dyn·mm	10	12	14	16	18	20	30	40	50	60	70	80	90	100

Table 4 shows the calculated probabilities of errors in the computation of the dynamic height of the sea surface with given depth of the reference surface. Errors in the computation of dynamic height for corresponding H values are given in the last line of the table for comparison. As we can see from the table, the probability of a zero error is small in absolute value. At the same time, the probability of errors not exceeding half the mistake in the computation of dynamic height is comparable with the probability of the zero error. The probability of larger errors is small. Moreover, the integral distribution of errors shows that in 90% of the cases the error in the computation of dynamic height does not exceed half the mistake. Therefore, in estimating the reliability of a dynamic chart, we can assume that the accuracy of dynamic height computation is half the mistake in the computation. In doing this we obviously reduce the accuracy of dynamic height computation somewhat, but consider, as it were, the influence of factors that were not taken into account earlier (the dependence of the error in the determination of sea water density on water temperature and salinity, the error in the computation of the temperature reduction correction, etc.).

10. ACCURACY OF CURRENT VELOCITY COMPUTATION BY THE DYNAMIC METHOD

The component of current velocity that is normal to a hydrological cross section is computed by the dynamic method from the difference between dynamic heights at two neighboring stations using formula

$$c = \frac{\Delta D}{2 \omega L \sin \varphi}. \qquad (2.33)$$

In this expression c is the component of current velocity that is normal to the cross section; $\Delta D = D_1 - D_2$ is the difference in the dynamic height of the isobaric surface at two neighboring hydrological stations; ω is the angular speed of rotation of the earth; D is the distance between stations; and φ is the geographical latitude of the area.

The absolute error in the computation of current velocity is estimated from the following expression:

$$dc = \frac{d\Delta D}{2\omega L \sin \varphi} + \frac{\Delta D}{2\omega L^2 \sin \varphi} dL. \quad (2.34)$$

It derives from the error in the computation of the difference in dynamic height and allowable errors in the measurement of the distance between neighboring hydrological stations. The latter derive from the accuracy in determination of the ship's location and also from errors in the measurement of distances on the chart. Having estimated both sources of errors, we find that the second term in formula (2.34) can be neglected.

In determining the ultimate error in the computation of current velocity by the dynamic method, let us assume that the errors in the computation of dynamic heights at two neighboring stations are equal in absolute magnitude, but opposite in sign. Then, according to (2.25), we have

$$dc = 2MH\, dv_t. \quad (2.35)$$

The designation $M = (2\omega L \sin\varphi)^{-1}$ is introduced in the above formula because there is a special table for coefficient M (Zubov, Chigirin, 1940).

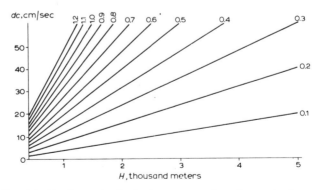

Fig. 14. Absolute mistake in the computation of current velocity by the dynamic method.

10. ACCURACY OF CURRENT VELOCITY COMPUTATION

Figure 14 is a graphical representation of the dependence of the absolute error in the computation of current velocity on the depth of the reference surface (or, more precisely, the distance between the reference and the isobaric surfaces). The ultimate error in the determination of specific volume is considered to be constant and equal to ± 0.02 σ_t units. Each sloping line corresponds to the M coefficient written above it. The graph shows clearly that in some cases the error in the computation of current velocity by the dynamic method may be rather large. It increases with increasing distance H between the reference and the given isobaric surfaces, and with decreasing geographical latitude of the area φ and distance between stations L.

The absolute mistake in the computation of current velocity as such cannot characterize the degree of accuracy of the result obtained. The same absolute error will confirm the great reliability of the picture of horizontal circulation when current velocities are high, whereas with low velocities it indicates that it is impossible to obtain an idea of currents by the dynamic method. Therefore, it is more convenient to use the relative mistake in the computation of current velocity when estimating the reliability of the computed current velocities. The relative mistake is

$$\frac{dc}{c} = \frac{d\Delta D}{\Delta D}. \qquad (2.36)$$

For the reasons given above, in this expression we do not consider the magnitude of the relative error in the measurement of the distances between neighboring hydrological stations.

Taking (2.25) into consideration, we obtain an expression for the relative mistake in the computation of current velocity by the dynamic method

$$\frac{dc}{c} = \frac{2H\,dv_t}{\Delta D}. \qquad (2.37)$$

This mistake is directly proportional to the distance H and to the mistake in the determination of the specific volume, and inversely proportional to the difference in dynamic heights at two neighboring hydrological stations. This relationship is presented graphically in Fig. 15. The sloping lines in the graph represent isolines of the relative mistake in the computation of current velocity. With a relative error of more than 50% the computed horizontal water circulation will obviously be completely unrealistic.

Now let us discuss the probability of error in the computation of current velocity.

The probability of error in the computation of the differences in the dynamic heights of an isobaric surface is estimated from formula (2.30), but now the statistical measure of accuracy is $\sqrt{2}$ times smaller than the statistical measure of the accuracy of dynamic height, because the number of terms in expression (2.28) has doubled. The distribution of errors in the computation of the difference in dynamic heights (Table 5) is similar to the distribution of errors in the computation of dynamic height. They obey the law of distribution of random values.

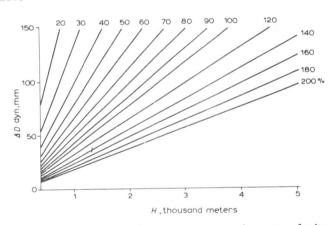

Fig. 15. Relative mistake in the computation of current velocity by the dynamic method depending on the difference in dynamic heights and the depth of the reference surface.

In analyzing the table and considering the arguments presented earlier, we may conclude that the actual error (in the sense of its probability of occurrence during

10. ACCURACY OF CURRENT VELOCITY COMPUTATION

Table 5

Probability of error in the computation of the differences in the dynamic heights of the sea surface, %

$[\delta]$, dyn·mm	500	600	700	800	900	1000	1500	2000	2500	3000	3500	4000	4500	5000
0	10.0	9.0	8.0	7.6	7.0	6.7	4.5	2.9	2.5	2.2	2.0	1.8	1.7	1.6
5	4.5	4.9	4.9	4.9	4.8	4.7	3.8	2.7	2.4	2.1	1.9	1.8	1.7	1.6
10	0.4	0.7	1.2	1.2	1.5	1.3	2.4	2.2	2.0	1.9	1.8	1.6	1.6	1.5
15			0.2	0.2	0.2	0.3	1.1	1.6	1.6	1.6	1.5	1.4	1.4	1.3
20							0.4	1.0	1.2	1.2	1.2	1.2	1.2	1.2
25							0.1	0.5	0.8	0.9	0.9	1.0	1.0	1.2
30								0.3	0.5	0.6	0.6	0.7	0.7	0.8
35								0.1	0.3	0.4	0.4	0.5	0.6	0.6
40									0.3	0.2	0.2	0.4	0.4	0.4
45									0.1	0.1	0.1	0.2	0.2	0.3
50											0.1	0.1	0.2	0.2
60 $[d\Delta D]$, dyn·mm	20	24	28	32	36	40	60	80	100	120	140	160	180	200

H, m

computation) in the difference in the dynamic heights of the isobaric surface is probably one-third of its maximum value, i.e., mistake. Errors with some given H values are presented in the last line of Table 5 for comparison.

To estimate the magnitude of the error that appears with a given probability $P(\vartheta)$ in computation of current velocity by the dynamic method, it is sufficient to compute the magnitude of the error with the same probability in the difference of dynamic heights and to substitute it in formula (2.34) or (2.35).

11. DISCUSSION OF RESULTS

In the preceding paragraphs we discussed in succession the accuracy with which initial dynamic method data are obtained and the errors that may arise in computing the dynamic heights of an isobaric surface and current velocity. We have established that the reliability of dynamic computations depends on the accuracy with which the vertical distribution of sea water density is determined at hydrological stations and on the distance between the reference surface and the isobaric surface at which current velocity is computed. There is no need to discuss the fact that the reliability of dynamic computation also depends to a similar extent on the density of the network of hydrological stations in the region under study and on the close timing of the survey.

We must note that in a very general analysis of the accuracy of the dynamic method it is impossible to give a strict estimate of the errors in computing individual values, since they largely depend on the methods used to obtain the initial data and on the accuracy of the measuring equipment. For this reason, the accuracy of each result obtained must be studied individually. The example analyzed earlier can be used as an approximate scheme for estimating the accuracy of a dynamic chart. The magnitude of the errors would differ in other cases, but there is no doubt that computational errors cannot be neglected without first being estimated.

The dynamic method yields sufficiently accurate results when horizontal circulation on the isobaric surface under

11. DISCUSSION OF RESULTS

examination is pronounced. The method may become unsuitable if horizontal circulation is weak. Therefore, it is very important in dynamic calculations to estimate the reliability of individual sections of a dynamic chart, proceeding from the accuracy with which initial data are obtained. To do this, a chart of the smoothed (within the limits of accuracy of dynamic height computation) dynamic relief of the isobaric surface must be constructed, in addition to an ordinary dynamic chart. The computed current will be reliable if the dynamic contours have a generally similar configuration on both charts. Thus, in the example studied earlier (Fig. 10), northerly currents in the western and eastern portions of the chart are unreliable, while the current in the center of the chart can be considered as fully reliable. Computational errors may distort its velocity, but in general the computed current must correspond to the actual current.

The error in the computation of dynamic height is somewhat affected by inaccuracies in the initial data for deep layers in the ocean, because the vertical distances between successive samples are great at great depths. To increase the accuracy of dynamic computations it is necessary to increase the accuracy of observations in the deep layers and avoid extrapolation of missing values. It is also desirable to increase the number of depth levels sampled to the maximum because the probability of the mutual compensation of errors increases with an increasing number of terms.

Earlier computations of currents by the dynamic method required only that the reference surface be located in the layer with minimum horizontal circulation, i.e., in the layer where current velocity is near zero. Retaining this requirement and considering the accuracy of the dynamic method, we must add that the reference surface should not be placed below a fully determined horizon, depending on the intensity of horizontal circulation on the isobaric surface under consideration. This statement will become clear if we recall that the error in the computation of dynamic height and of current velocity is directly dependent on the distance H between the reference surface and a given isobaric surface, or directly proportional to the depth of the reference surface when current velocities at the sea surface are computed. If horizontal circulation is weak and the reference surface is located at a very great depth, the error in the computed difference of dynamic heights may become comparable to the actual differences in dynamic heights or even exceed them. In this case it makes sense to sacrifice absolute dynamic computation in favor of exact computations and to place the reference surface during dynamic analysis at a depth where computational errors will not influence the reliability of the result significantly. In the next chapter we shall describe a method of increasing the accuracy of dynamic computation in cases where the "zero" surface is determined with sufficient reliability, but is located at too great a depth.

In stating that selection of the reference surface for dynamic computation requires prior knowledge of the intensity of the horizontal circulation at the depth under investigation, we place the investigator trying to follow our advice in a vicious circle of unsolved problems, since the purpose of the dynamic method is to obtain an idea of horizontal water circulation. Let us therefore indicate a possible way of overcoming these difficulties.

Having determined the "zero" surface, we must estimate the error in the computation of dynamic heights relative to the selected "zero" surface from its depth H and from the accuracy of the determination of the specific volume dv_t. Having determined the limits of the existing differences in dynamic heights, we must estimate the order of magnitude of the error in the computation of current velocity from formula (2.27). If the latter exceeds 50% it is better to give up dynamic computation, because the layer of no motion is too deep and the probability of large computational errors makes any result dubious. For example, it has been established that current velocity is insignificantly small at a depth $H = 3000$ m. Assuming that $dv_t = \pm 0.02$, let us estimate the mistake in the computation of the difference in dynamic heights

$$d\Delta D = 2H\, dv_t = \pm 120 \text{ dyn} \cdot \text{mm}.$$

The probability of a maximum error is small; therefore let us assume that the actual error in the computation of the difference in dynamic heights is ± 40 dyn·mm. If the actually observed difference in the dynamic heights of the sea surface is of the order of 50-70 dyn·mm, the error in the computation of current velocity will be of the order of 60-80%. Here it is better to relinquish absolute dynamic computation and perform a dynamic analysis relative to the reference surface $H < 3000$ m.

We have already mentioned that the necessity to examine the accuracy of the dynamic method arose because the gradient component of the velocity of a wind-driven current penetrates to an infinitely great depth. Current velocity cannot be computed by the dynamic method relative to a very deep reference surface without risking distortion of the result by the accumulation of computational errors. Thus the problem that arises is selection of the optimum depth: that at which, on the one hand, current velocity is sufficiently small to be neglected, but which on the other hand is not so great that the current computed by the dynamic method will be unreliable. This problem is examined in the following chapter.

CONCLUSIONS

1. Evaluation of the accuracy of dynamic computation shows that in some cases computational errors may completely distort the result. Evaluation of the reliability of the result must be an obligatory part of dynamic analysis.

2. The extent to which the pattern of currents computed by the dynamic method approximates actual conditions depends on the depth of the reference surface and on the intensity of horizontal circulation at the depth under study, and also on the scale division (precision) of the measuring equipment, the number of depth levels sampled, the density of the network of hydrological stations, etc. Relative computational errors are larger, the greater the distance to the reference surface and the weaker the circulation at the given depth.

3. The reference surface used for dynamic computation cannot be located at very great depth because then computational errors increase greatly. The maximum allowable depth of the reference surface, from the viewpoint of the accuracy of the dynamic method, can be estimated on the basis of the results obtained in this chapter.

4. In some cases when computational errors are large and there is no confidence in the reliability of the computation, it is convenient to compute dynamic heights relative to a shallower reference surface and to sacrifice the computed absolute velocities in favor of more exact calculation.

CHAPTER III

Investigation of the Vertical Velocity Distribution of a Wind-Driven Current in a Deep Sea Using a Density Model

12. STATEMENT OF THE PROBLEM

Great progress has been made recently in the theory of stationary wind-driven currents in a deep inhomogeneous sea. V. B. Shtokman's method of full currents (1949, 1951), which studies the total vertical transport of water within the baroclinic layer, has been especially fruitful. In this method the wind field over the sea or the distribution of atmospheric pressure at the sea surface is assumed to be known. The factor sought is the function of the full current or of the component of the full current along the coordinate axes.

Since a full current as such does not give an idea of the vertical distribution of current velocity in a sea, density models are ordinarily used to compute current velocities at individual depths (Reid, 1948; Shtokman, 1950, 1951; Kitkin, 1953). Without examining the various density models, let us note that the most generalized model is that of V. B. Shtokman. By making the sea water density field conform to some model, it is possible to compute from the wind field the readjustment of density in the sea as a result of wind action on the sea surface and then, from the density distribution, the gradient component of current velocity in addition to full currents (Shtokman, 1950). Summation of the gradient component with the pure drift component of current velocity yields the velocity of the wind-driven current at individual depths in the sea.

There are now several known methods for computing full currents in the sea from wind or atmospheric pressure. Most of these are based on the solution of the biharmonic equation of V. B. Shtokman that relates the function of the full current to the curl in the tangential wind stress field. To solve this equation we must know the wind distribution over the entire sea and boundary conditions along the margin of the sea. The latter are easily formulated at the shoreline and reduce either to conditions of inpenetrability and adhesion for a full current, or to conditions of impenetrability and sliding. It is difficult to formulate boundary conditions at the liquid limits of open seas or individual areas in the ocean. This complicates the problem considerably.

An analytical solution of the Shtokman biharmonic equation can be obtained only for an idealized sea of regular geometric configuration. Numerical integration of the biharmonic equation is normally performed in computing full currents in an actual sea. Modern computation techniques make it possible to compute the field of full currents in a sea of any form and with any given wind field quickly and easily. Positive results have been obtained from experimental models of full currents based on the analogy between these currents and the deflection of a thin plate that is fixed along its margin. The distribution of the function of the full current over the water area can be rapidly determined from a model of the sea (Shtokman, 1953b; Fedorov, 1956).

Along with the successful development of the theory of wind-driven ocean currents, considerable progress has been made in recent years in the direct measurement of current velocities at sea. Modern techniques make it possible to measure surface currents over large ocean areas with sufficient accuracy. One of the foremost instruments is the electromagnetic current meter that allows measurement of current velocity from a ship in motion.

The accumulation of data from direct measurement of surface current velocities will probably make it possible in the near future to construct surface current charts that are averaged over a sufficiently long time interval. As we know, averaged current velocities can be identified with sufficient accuracy with stationary currents. Therefore it would be interesting to develop a method of computing current velocities in the entire water column, and estimate the depth of penetration of a wind-driven ocean current using theory and the results of direct measurements of surface current velocities.

In doing this we should not forget that we are only considering the stationary state. In the computation we must use the velocities of a surface current that approximates a stationary current and, in practical application, the steadiest currents or currents that are averaged over a long time interval (season, year). When estimating the depth of penetration of a current, it is very important that the velocities of the surface current and the full current correspond to each other, i.e., describe the same dynamic state of water masses.

13. METHOD OF COMPUTING DEEP SEA CURRENTS FROM THE SURFACE CURRENT AND ATMOSPHERIC PRESSURE GRADIENT

As we have shown, a wind-induced current in the sea represents the sum of a pure drift component directly generated by the entraining action of the wind and a gradient component due to the tidal effect of wind and to the readjustment of the water density field.

The gradient component of the velocity of a steady current in a deep inhomogeneous sea is sufficiently accurately determined by the dynamic method, which assumes the existence of a horizontal "zero" isobaric surface at some depth H and neglects frictional forces in a stationary gradient current. That these assumptions are permissible has been demonstrated earlier. The strict requirement that the "zero" surface be horizontal can be removed and we can assume that it is of arbitrary shape. Subsequent formulas will include additional terms that will depend on the derivatives of the depth of the "zero" surface along the coordinate axes X and Y. In identifying the "zero" surface with a horizontal isobaric surface, we assume that in a region with an identical vertical structure of water masses the "zero" surface has a small slope; we neglect terms that contain $\frac{\partial H}{\partial x}$ and $\frac{\partial H}{\partial y}$ because they are relatively small.

According to the dynamic method, the expressions for the horizontal components of the gradient component of current velocity are written in the following form:*

$$u_g = \frac{g}{\Omega \bar{\rho}} \int_0^z \frac{\partial \rho}{\partial y} dz;$$

$$v_g = -\frac{g}{\Omega \bar{\rho}} \int_0^z \frac{\partial \rho}{\partial x} dz.$$

(3.1)

In expressions (3.1) u_g and v_g are the components of the gradient current along the Cartesian coordinate axes X and Y; g is the acceleration of gravity; ρ is the density of sea water; $\bar{\rho}$ is its average value, assumed to be constant; $\Omega = 2\omega \sin \varphi$ is the coriolis parameter, in which ω is the angular speed of rotation of the earth and φ is the geographical latitude of the area. The origin of the coordinates is located on the horizontal "zero" isobaric surface, the Z axis is directed vertically upward, and $z = H$ at the sea surface.

According to the density model of V. B. Shtokman, the deviation of density $\rho(x, y, z)$ from its constant value at the "zero" surface, i.e., the lower boundary of the baroclinic layer [let us designate this density value as $\rho(0)$], is given as the product of two functions, one of which depends only on the vertical coordinate and the other only on the horizontal coordinate (Shtokman, 1950, 1951)

$$\rho(0) - \rho(x, y, z) = \delta(z) \cdot f(x, y).$$

(3.2)

Knowing the distribution of density along the vertical at point $A(x_A, y_A)$ and assuming that $f(x_A, y_A) = 1$, from (3.2) we obtain

$$\delta(z) = \rho_A(0) - \rho_A(z),$$

(3.3)

---

*Here and henceforth we shall consider the Northern Hemisphere. The signs in expressions (3.1) change to the opposite for the Southern Hemisphere.

where $\rho_A(z)$ is the vertical distribution of density at point A.

Thus we have determined the function $\delta(z)$ that represents the difference in density between the "zero" surface and depth z. Computation of density and then of the gradient current reduces to the determination of function $f(x, y)$, called the Function of Influence by V. B. Shtokman.

The density model described by expression (3.2) is based on the constancy of density at the lower boundary of the baroclinic layer. Moreover, it implies that the vertical density distribution curves are identical because, according to (3.2), the difference in densities at the lower boundary of the baroclinic layer and at depth z on any vertical is obtained by multiplying the corresponding difference on the initial vertical by the function of influence, which only depends on the horizontal coordinates. The selection of the initial vertical is arbitrary.

The similarity of the vertical distribution of sea water density on which the model is based represents one of the most essential features of the actual density distribution in a sea. It is easy to show that it is analogous to the similarity of T-S curves. Analysis of observational data shows that the similarity of T-S curves, i.e., the identity of the vertical structure of water masses, is retained over huge ocean and sea areas.

To use V. B. Shtokman's model we must evidently divide the ocean or sea into regions by T-S curves. Many oceans and seas have already been divided into regions in this manner. The number of regions with an identical vertical structure of water masses is small. For each such region it is sufficient to know from observation the vertical density distribution along one vertical only, and consider it the initial vertical in a region with a constant water mass structure. We shall show below that the errors in gradient current velocity resulting from the arbitrary selection of the initial vertical are insignificant, since the effect of the vertical stratification of sea water considered by the model changes very little in the horizontal direction.

Shtokman's model, which is the most generalized of the existing models, cannot be applied to a stratified current. The constancy of the function of influence along the vertical implies that the velocity of the gradient current changes with depth in absolute magnitude only, remaining constant in direction. Consequently, the boundary between differently directed branches of a circulation pattern can only be vertical, according to this model. The gradient current in a deep sea must actually be such, far from shore.

By substituting (3.2) in (3.1) we obtain an expression for the horizontal components of the gradient current

$$u_g = \frac{-g}{\Omega\bar{\rho}} \frac{\partial f}{\partial y} \int_0^z \delta(z)\,dz;$$
$$v_g = \frac{g}{\Omega\bar{\rho}} \frac{\partial f}{\partial x} \int_0^z \delta(z)\,dz.$$
(3.4)

The horizontal components of the velocity of a pure wind-driven current, designated as u_d and v_d, are defined in a deep sea by expressions derived from the Ekman theory. With a wind of random direction they have the following form:

$$u_d = \frac{e^{-a(H-z)}}{2a\mu}[(T_x + T_y)\cos a(H-z) - (T_x - T_y)\sin a(H-z)];$$
$$v_d = \frac{e^{-a(H-z)}}{2a\mu}[(T_x + T_y)\sin a(H-z) + (T_x - T_y)\cos a(H-z)].$$
(3.5)

T_x and T_y in expressions (3.5) are the components of tangential wind stress along the coordinate axes X and Y, respectively; μ is the coefficient of vertical turbulent exchange of the amount of motion that we consider to be independent of the vertical coordinate, but to vary in the horizontal directions; and a is a quantity given by formula

$$a = \sqrt{\frac{\Omega\bar{\rho}}{2\mu}}.$$
(3.6)

13. METHOD OF COMPUTING DEEP SEA CURRENTS

Summing formulas (3.4) and (3.5), we obtain the following expression for the horizontal velocity components of the current sought:

$$u = \frac{-g}{\Omega \bar{\rho}} \frac{\partial f}{\partial y} \int_0^z \delta(z)\,dz + \frac{e^{-a(H-z)}}{2a\mu} [(T_x + T_y)\cos a(H-z) - (T_x - T_y)\sin a(H-z)];$$
$$v = \frac{g}{\Omega \bar{\rho}} \frac{\partial f}{\partial x} \int_0^z \delta(z)\,dz + \frac{e^{-a(H-z)}}{2a\mu} [(T_x + T_y)\sin a(H-z) + (T_x - T_y)\cos a(H-z)].$$
(3.7)

Assuming that $z = H$ in formulas (3.7), we obtain expressions for the horizontal velocity components of the surface current

$$u(H) = \frac{-g}{\Omega \bar{\rho}} \frac{\partial f}{\partial y} \int_0^H \delta(z)\,dz + \frac{1}{2a\mu}(T_x + T_y);$$
$$v(H) = \frac{g}{\Omega \bar{\rho}} \frac{\partial f}{\partial x} \int_0^H \delta(z)\,dz + \frac{1}{2a\mu}(T_x - T_y).$$
(3.8)

The only unknown values in relationships (3.8) are the partial derivatives of the function of influence $\frac{\partial f}{\partial x}$ and $\frac{\partial f}{\partial y}$, since we know the velocity of the current at the sea surface from observation. By determining the unknowns from (3.8) and substituting the result in expression (3.7), we obtain the following formulas that describe the vertical distribution of current velocity

$$u(z) = u(H)\Phi(z) - \frac{\Phi(z)}{2a\mu}(T_x + T_y) + \frac{e^{-a(H-z)}}{2a\mu} \times [(T_x + T_y)\cos a(H-z) - (T_x - T_y)\sin a(H-z)];$$
$$v(z) = v(H)\Phi(z) + \frac{\Phi(z)}{2a\mu}(T_x - T_y) + \frac{e^{-a(H-z)}}{2a\mu} \times [(T_x + T_y)\sin a(H-z) + (T_x - T_y)\cos a(H-z)],$$
(3.9)

where we adopt the following designation

$$\Phi(z) = \frac{\int_0^z \delta(z)\,dz}{\int_0^H \delta(z)\,dz}. \qquad (3.10)$$

The components of tangential wind stress, T_x and T_y, entering expression (3.9) are conveniently expressed by the component of the atmospheric pressure gradient. To do this, let us write the equation of steady motion in the atmosphere in the form

$$\mu_1 \frac{\partial^2 u_1}{\partial z^2} + \Omega \bar{\rho}_1 v_1 = \frac{\partial p}{\partial x};$$

$$\mu_1 \frac{\partial^2 v_1}{\partial z^2} - \Omega \bar{\rho}_1 u_1 = \frac{\partial p}{\partial y}, \qquad (3.11)$$

where u_1 and v_1 are the components of wind velocity along the X and Y axes; p is atmospheric pressure; μ_1 is the coefficient of the vertical turbulent exchange of the amount of motion in the atmosphere that is independent of the vertical coordinate; and $\bar{\rho}_1$ is the average air density. The system of coordinates is the same and $z = H$ at the sea surface.

If we assume, as is ordinarily done in dynamic meteorology, that the horizontal pressure gradient does not vary with height, then instead of (3.11) we can write

$$\mu_1 \frac{\partial^2 u_1}{\partial z^2} + \Omega \bar{\rho}_1 v_1 = \Omega \bar{\rho}_1 v_{g1};$$

$$\mu_1 \frac{\partial^2 v_1}{\partial z^2} - \Omega \bar{\rho}_1 u_1 = -\Omega \bar{\rho}_1 u_{g1}. \qquad (3.12)$$

In equations (3.12) u_{g1} and v_{g1} are the geostrophic wind velocity components given by expressions

$$u_{g1} = -\frac{1}{\Omega \bar{\rho}_1} \frac{\partial p}{\partial y};$$

$$v_{g1} = \frac{1}{\Omega \bar{\rho}_1} \frac{\partial p}{\partial x}. \qquad (3.13)$$

13. METHOD OF COMPUTING DEEP SEA CURRENTS

Multiplying the second equation in (3.12) by $i = \sqrt{-1}$ and adding it to the first, we obtain the equation of motion in the atmosphere in the complex form

$$-\frac{1}{q}\frac{\partial^2 \mathfrak{v}_1}{\partial z^2} + \mathfrak{v}_1 = \mathfrak{v}_{g1}, \qquad (3.14)$$

that contains the following designations:

$$\mathfrak{v}_1 = u_1 + iv_1; \qquad (3.15)$$
$$\mathfrak{v}_{g1} = u_{g1} + iv_{g1}; \qquad (3.16)$$
$$q^2 = \frac{\Omega \bar{\rho}_1}{\mu_1} i. \qquad (3.17)$$

Assuming that there is no air motion at the sea surface and that wind is geostrophic at great heights, we can write the boundary conditions in the form:

$$(\mathfrak{v}_1)_{z=H} = 0; \qquad (3.18)$$
$$(\mathfrak{v}_1)_{z\to\infty} = \mathfrak{v}_{g1}. \qquad (3.19)$$

The solution of equation (3.14) that satisfies boundary conditions (3.18) and (3.19) has the form

$$\mathfrak{v}_1 = \mathfrak{v}_{g1}(1 - e^{-qz}). \qquad (3.20)$$

In its complex form the relationship between tangential wind stress and the vertical distribution of wind velocity is:

$$\overline{T} = \mu_1 \left(\frac{\partial \mathfrak{v}_1}{\partial z}\right)_{z=H}, \qquad (3.21)$$

where

$$\overline{T} = T_x + iT_y. \qquad (3.22)$$

Substituting (3.20) in (3.21) we obtain

$$\overline{T} = q\mu_1 \mathfrak{v}_{g1}. \qquad (3.23)$$

III. INVESTIGATION OF THE VERTICAL VELOCITY

Or, by separating the actual from the imaginary part in the last expression, we obtain the following final expression for the components of tangential wind stress:

$$T_x = \sqrt{\frac{\mu_1}{2\Omega\bar{\rho}_1}}\left(\frac{\partial p}{\partial x} + \frac{\partial p}{\partial y}\right);$$

$$T_y = \sqrt{\frac{\mu_1}{2\Omega\bar{\rho}_1}}\left(\frac{\partial p}{\partial x} - \frac{\partial p}{\partial y}\right).$$

(3.24)

According to formulas (3.24), the wind at the sea surface deflects by an angle of 45° to the left of the direction of isobars. Actually, observations show that this angle is somewhat smaller than 45°. The discrepancy between computation and observation is attributable to the fact that the actual coefficient of vertical exchange of the amount of motion in the atmosphere, which was assumed by us to be independent of the vertical coordinate, varies with height. But this discrepancy is largely canceled out by the fact that in computing the pure drift component of current velocity we also assume that the coefficient of vertical exchange of the amount of motion in the sea is independent of the vertical coordinate. As a result, the pure wind-driven current at the sea surface deflects by an angle of 45° to the right of the wind direction and is oriented along the isobar. The same result was obtained earlier by Exner (1912) and apparently agrees well with actual conditions.

Now substituting formula (3.24) in (3.9), we obtain an expression for determining the velocity of the deep current from the surface current

$$u(z) = \left[u(H) + k\frac{\partial p}{\partial y}\right]\Phi(z) + ke^{-a(H-z)}\left[\frac{\partial p}{\partial x}\sin a(H-z) - \frac{\partial p}{\partial y}\cos a(H-z)\right];$$

$$v(z) = \left[v(H) - k\frac{\partial p}{\partial x}\right]\Phi(z) - ke^{-a(H-z)}\left[\frac{\partial p}{\partial x}\cos a(H-z) + \frac{\partial p}{\partial y}\sin a(H-z)\right],$$

(3.25)

where coefficient k is given by the following formula:

$$k = \frac{1}{\Omega}\sqrt{\frac{\mu_1}{\mu\bar{\rho}\rho_1}}. \qquad (3.26)$$

Expressions (3.25) define current velocities at individual depths depending on current velocity at the sea surface, the distribution of atmospheric pressure, the vertical stratification of sea water, and also on the magnitude of k and a, functions of the not yet determined quantities μ and μ_1. Therefore, let us now determine the coefficients of vertical exchange of the amount of motion in the atmosphere and in the sea. Let us assume that they are independent of the vertical coordinate, but vary along horizontal directions.

From equations (3.24) we obtain

$$T = \sqrt{\frac{\mu_1}{\Omega\bar{\rho}\rho_1}} \cdot \frac{\partial p}{\partial n}, \qquad (3.27)$$

where T is the absolute value of tangential wind friction and n is the direction of the normal to the isobar.

Tangential wind stress is ordinarily calculated by the following well-known formula:

$$T = \gamma w_1^2. \qquad (3.28)$$

Here γ is a constant factor equal to $3.25 \cdot 10^{-6}$ g/cm^3 and w_1 is wind velocity at anemometer level, which in its turn can be related to geostrophic wind velocity by the formula

$$w_1 = \lambda w_{g1}. \qquad (3.29)$$

In formula (3.29) λ is some function of geostrophic wind velocity w_{g1}, of the geographic latitude of the area φ, and of the atmospheric stratification index ε (Drogaitsev, 1954). Quantity ε is associated with the lapse rate θ by the empirical relationship:

$$\varepsilon = 0.17789 - 0.13480\theta - 0.00975\theta^2. \qquad (3.30)$$

Table 6 gives (by months) the index of atmospheric stratification in the temperate latitudes of the Northern Hemisphere, obtained from the analysis of a great number of observations.

Table 6

Atmospheric stratification coefficient over the open sea at temperate latitudes in the Northern Hemisphere (according to D. A. Drogaitsev)

Months											
I	II	III	IV	V	VI	VII	VIII	IX	X	XI	XII
0.05	0.05	0.07	0.10	0.14	0.16	0.17	0.16	0.12	0.07	0.05	0.05

Geostrophic wind velocity entering expression (3.29) is determined, according to (3.13), by the formula

$$w_{g1} = \frac{1}{\Omega \rho_1} \frac{\partial p}{\partial n}. \tag{3.31}$$

Combining equations (3.28), (3.29), and (3.31) we obtain

$$T = \frac{\gamma \lambda^2}{\Omega^2 \rho_1^2} \left(\frac{\partial p}{\partial n} \right)^2. \tag{3.32}$$

Comparing expressions (3.27) and (3.32) for tangential wind stress, we obtain the following formula for the coefficient of vertical turbulent exchange of the amount of motion in the atmosphere:

$$\mu_1 = \frac{\gamma^2 \lambda^4}{\Omega^3 \rho_1^3} \left(\frac{\partial p}{\partial n} \right)^2. \tag{3.33}$$

If we compute the values of the coefficient of turbulent exchange in the atmosphere by formula (3.33), they will be lower by about an order of magnitude than the values computed by methods used on land (Laikhtman and Chudnovskii, 1949). This discrepancy can probably be explained

13. METHOD OF COMPUTING DEEP SEA CURRENTS

by the fact that the sea surface has less influence on the development of turbulent processes in the atmosphere than does the land surface.

Let us now determine the coefficient of vertical turbulent exchange of the amount of motion in the sea. Proceeding from consideration of the dimensionality of coefficient μ (g/cm·sec), it is natural to assume that it depends on wind velocity w_1, the coriolis parameter Ω, and sea water density $\bar{\rho}$. The only combination of w_1, Ω, and $\bar{\rho}$ that would have the dimensionality of coefficient μ is the following expression:

$$\mu = C\bar{\rho}\frac{w_1^2}{\Omega}, \qquad (3.34)$$

where C is some constant dimensionless value. Thus the coefficient of vertical turbulent friction in the sea is directly proportional to sea water density and the square of wind velocity, and inversely proportional to the coriolis parameter. The same result can be obtained by using the classical theory of Ekman. Moreover, the Ekman theory makes it possible to determine the dimensionless coefficient C that enters into expression (3.34).

According to Ekman, the modulus of the velocity of a pure wind-driven surface current is defined by the expression

$$w_d = \frac{T}{a\mu\sqrt{2}}. \qquad (3.35)$$

By substituting expression (3.28) in expression (3.35) and dividing both parts of the result by wind velocity, we obtain the following expression for the ratio of the velocity of the pure wind-driven current at the sea surface to wind velocity:

$$\frac{w_d}{w_1} = \frac{\gamma w_1}{a\mu\sqrt{2}}. \qquad (3.36)$$

As a result of analysis of a large number of measurements of surface current velocity and wind velocity, many authors have obtained the following relationship between

III. INVESTIGATION OF THE VERTICAL VELOCITY

the velocity of a pure wind-driven surface current and wind velocity (Shuleikin, 1953):

$$\frac{w_d}{w_1} = \frac{0.0127}{\sqrt{\sin \varphi}}. \tag{3.37}$$

Considering formulas (3.6), (3.29), (3.31), (3.36), and (3.37), we obtain the following final expression for the coefficient of vertical turbulent exchange of the amount of motion in the sea:

$$\mu = \frac{\gamma^2 \lambda^2 \sin \varphi}{1.6130^2 \overline{\rho} \rho_1^2} \left(\frac{\partial p}{\partial n}\right)^2 \cdot 10^4. \tag{3.38}$$

Substituting the expressions for the coefficients of vertical exchange in the atmosphere (3.33) and in the sea (3.38) in (3.26) and (3.6), we obtain the final formulas for the sought coefficients k and a:

$$k = \frac{0.0127\lambda}{2\omega \overline{\rho}_1 \sin^{3/2}\varphi}; \tag{3.39}$$

$$a = \frac{0.0358\omega^2 \overline{\rho\rho}_1 \sin^{3/2}\varphi}{\gamma\lambda \frac{\partial p}{\partial n}}. \tag{3.40}$$

The last two formulas contain function λ that depends, as we mentioned earlier, on the velocity of the geostrophic wind, the geographical location of the area, and the atmospheric stratification index. Geostrophic wind velocity in its turn depends on the horizontal atmospheric pressure gradient and the latitude of the area. Thus, λ is a function of the horizontal atmospheric pressure gradient $\frac{\partial p}{\partial n}$, the geographical latitude of the area φ, and the atmospheric stratification index ε. A rather complex functional dependence of λ on w_{g1}, φ, and ε is presented in a paper of D. A. Drogaitsev (1954). There are also special tables for determining λ. For our purpose it is more convenient, instead of tables of function λ, to use tables giving direct values for the k and a coefficients relative to $\frac{\partial p}{\partial n}, \varphi$, and ε. Tables 7 and 8 represent such

Table 7

Coefficient $k \cdot 10^{-5}$

$\frac{\partial p}{\partial n} \cdot 10^5$, dyn/cm³

φ, degree	1	5	10	15	20	25	30	35	40	45	50

$\epsilon = 0.05$

10	6.51	6.42	6.24								
20	2.68	2.48	2.41	2.34	2.34	2.31					
30	1.58	1.46	1.41	1.39	1.37	1.35	1.35	1.33			
40	1.09	1.02	0.99	0.98	0.96	0.95	0.95	0.94	0.94	0.93	0.92
50	0.86	0.80	0.77	0.76	0.75	0.74	0.74	0.73	0.73	0.72	0.71
60	0.72	0.67	0.65	0.63	0.63	0.62	0.61	0.61	0.61	0.61	0.61
70	0.64	0.59	0.57	0.56	0.56	0.55	0.55	0.54	0.54	0.53	0.53
80	0.60	0.56	0.54	0.53	0.52	0.52	0.51	0.51	0.50	0.50	0.50

$\epsilon = 0.10$

10	5.58	4.84	4.56	4.47							
20	2.24	1.94	1.84	1.78	1.74	1.71					
30	1.37	1.20	1.12	1.08	1.06	1.04	1.02	1.01			
40	0.95	0.84	0.81	0.78	0.75	0.74	0.73	0.72	0.71	0.70	0.70
50	0.77	0.68	0.62	0.61	0.60	0.59	0.58	0.57	0.56	0.56	0.55
60	0.66	0.56	0.54	0.52	0.50	0.50	0.48	0.48	0.47	0.47	0.47
70	0.58	0.50	0.47	0.46	0.44	0.44	0.44	0.43	0.43	0.42	0.42
80	0.54	0.48	0.45	0.44	0.43	0.42	0.41	0.41	0.40	0.40	0.39

$\epsilon = 0.15$

10	4.75	3.91	3.63	3.54							
20	2.01	1.64	1.54	1.54	1.44	1.41	1.37				
30	1.24	1.02	0.95	0.91	0.87	0.86	0.84	0.82			
40	0.87	0.74	0.69	0.66	0.64	0.62	0.60	0.60	0.59	0.58	0.57
50	0.73	0.61	0.53	0.53	0.52	0.50	0.49	0.47	0.46	0.46	0.46
60	0.62	0.51	0.47	0.44	0.43	0.42	0.41	0.41	0.40	0.39	0.39
70	0.55	0.45	0.42	0.40	0.39	0.38	0.37	0.37	0.36	0.36	0.35
80	0.52	0.43	0.40	0.38	0.37	0.36	0.35	0.35	0.34	0.34	0.33

$\epsilon = 0.20$

10	4.28	3.35	3.07	2.98							
20	2.46	1.94	1.76	1.63	1.58	1.54					
30	1.18	0.95	0.86	0.80	0.78	0.74	0.72	0.70			
40	0.84	0.69	0.62	0.58	0.57	0.55	0.53	0.52			
50	0.73	0.57	0.50	0.48	0.47	0.45	0.43	0.42	0.41	0.41	0.41
60	0.62	0.48	0.43	0.41	0.39	0.38	0.37	0.37	0.36	0.35	0.35
70	0.55	0.43	0.39	0.38	0.35	0.34	0.33	0.33	0.32	0.32	0.31
80	0.52	0.41	0.38	0.35	0.34	0.33	0.32	0.31	0.30	0.30	0.29

Table 8

Coefficient $a \cdot 10^4$

$$\frac{\partial p}{\partial n} \cdot 10^5, \text{ dyn/cm}^3$$

ψ, degree	1	5	10	15	20	25	30	35	40	45	50
\multicolumn{12}{c}{$\epsilon = 0.05$}
10	7.29	1.59	0.80	0.55							
20	19.00	4.11	2.11	1.44	1.09	0.86					
30	32.20	6.97	3.62	2.45	1.86	1.51	1.26	1.10			
40	46.70	10.00	5.16	3.49	2.65	2.15	1.80	1.56	1.36	1.22	1.10
50	58.5	12.75	6.62	4.46	3.40	2.74	2.30	2.00	1.76	1.57	1.41
60	70.5	15.1	7.76	5.38	4.03	3.26	2.76	2.36	2.10	1.86	1.68
70	78.8	17.1	8.88	6.00	4.56	3.70	3.08	2.67	2.37	2.11	1.90
80	84.4	18.1	9.40	6.35	4.83	3.91	3.31	2.87	2.51	2.23	2.01
\multicolumn{12}{c}{$\epsilon = 0.10$}
10	9.12	2.11	1.05	0.76							
20	22.7	5.24	2.76	1.91	1.46	1.20					
30	37.2	8.50	4.55	3.14	2.39	1.95	1.61	1.44			
40	53.7	12.05	6.31	4.35	3.38	2.76	2.34	2.04	1.78	1.56	1.45
50	64.4	15.00	8.18	5.55	4.25	3.45	2.92	2.56	2.28	2.02	1.82
60	76.6	18.05	9.43	6.58	5.05	4.08	3.51	3.01	2.68	2.39	2.14
70	86.5	20.1	10.70	7.20	5.67	4.63	3.85	3.41	2.99	2.64	2.43
80	93.0	21.2	11.2	7.75	5.90	4.80	4.13	3.61	3.15	2.81	2.56
\multicolumn{12}{c}{$\epsilon = 0.15$}
10	10.8	2.61	1.40	0.96							
20	25.0	6.20	3.30	2.35	1.80	1.49					
30	41.3	9.95	5.36	3.72	2.92	2.37	2.03	1.78			
40	58.5	13.7	7.40	5.13	4.00	3.26	2.83	2.44	2.13	1.98	1.78
50	67.8	16.6	9.32	6.40	4.88	4.07	3.45	3.08	2.76	2.45	2.22
60	80.5	20.0	10.80	7.69	5.88	4.79	4.16	3.56	3.18	2.76	2.61
70	91.2	22.4	12.20	8.40	6.53	5.43	4.52	3.96	3.54	3.14	2.88
80	96.6	23.2	12.6	8.86	6.89	5.62	4.86	4.25	3.72	3.30	3.04
\multicolumn{12}{c}{$\epsilon = 0.20$}
10	11.96	3.06	1.66	1.15							
20	27.2	6.91	3.80	2.69	2.11	1.74					
30	43.4	10.75	5.98	4.27	2.92	2.76	2.36	2.08			
40	60.2	14.8	8.15	5.80	4.45	3.72	3.18	2.79	2.50	2.28	2.06
50	69.6	18.5	9.97	7.05	5.41	4.51	3.76	3.46	3.10	2.76	2.48
60	81.5	21.0	11.8	8.31	6.50	5.42	4.62	3.97	3.55	3.24	2.91
70	92.1	23.8	13.0	9.03	7.20	6.00	5.00	4.38	4.01	3.57	3.29
80	97.6	24.7	13.5	9.70	7.42	6.18	5.38	4.71	4.12	3.84	3.45

13. METHOD OF COMPUTING DEEP SEA CURRENTS

tables with four values of the atmospheric stratification index ($\varepsilon = 0.05, 0.10, 0.15$, and 0.20). For other values of ε the coefficients sought may be found by interpolation.

Using an example, let us now compute the vertical distribution of current velocity from formulas (3.25). In this example we shall compute the average monthly depths of currents for October. Figure 16 represents a section of a monthly mean atmospheric pressure chart with isobars drawn every 1 mb. The coordinate axes conform to the selected system of coordinates: the X axis is directed to the east and the Y axis to the north. The vertical distribution of current velocity is computed for point B. The current velocity at the same point at the sea surface is known from observation and its components are directed along the coordinate axes: $u(H) = 15.4$ cm/sec and $v(H) = 5.4$ cm/sec. The vertical distribution of water density at points M and N is known, and is given in Table 9.

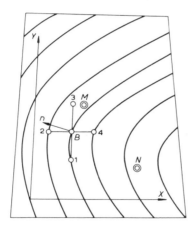

Fig. 16. Section of a mean monthly atmospheric pressure chart: n—direction of the normal to the isobar; 1, 2, 3, 4—points used to compute the pressure gradients.

For computation purposes it is sufficient to know from observation the density distribution at one vertical only. The second vertical is given solely for estimating the variability of the effect of the vertical stratification of sea water in the horizontal direction.

With the distribution of points and coordinate axes given in Fig. 16, the horizontal atmospheric pressure gradient is computed from the following differential equations:

$$\frac{\partial p}{\partial x} = \frac{1}{2\Delta x}(p_4 - p_2);$$

$$\frac{\partial p}{\partial y} = \frac{1}{2\Delta y}(p_3 - p_1).$$

(3.41)

Table 9

Vertical density distribution σ_t at stations M and N

Depth, m	0	10	25	50	75	100	200	300
σ_t, station M	26.04	26.05	26.05	26.64	26.70	26.72	26.90	27.02
σ_t, station N	26.03	26.03	26.05	26.40	26.72	26.76	26.92	27.07
Depth, m	400	500	750	1000	1500	2000	2500	3000
σ_t, station M	27.15	27.26	27.36	27.42	27.59	27.71	27.77	27.80
σ_t, station N	27.17	27.20	27.37	27.44	27.55	27.58	27.65	27.72

In our example, the mean monthly atmospheric pressure at these points is:

$p_1 = 1004.0$ mb $\qquad p_3 = 1004.6$ mb
$p_2 = 1006.0$ mb $\qquad p_4 = 1003.0$ mb

The finite distance increments are $2\Delta x = 3°$ along the parallel and $2\Delta y = 2°$ along the meridian. Considering that 1 mb = 1000 dyn/cm^2, we obtain

$$\frac{\partial p}{\partial x} = -11.1 \cdot 10^{-5} \text{ dyn/cm}^3; \frac{\partial p}{\partial y} = 2.9 \cdot 10^{-5} \text{ dyn/cm}^3;$$

$$\frac{\partial p}{\partial n} = 11.4 \cdot 10^{-5} \text{ dyn/cm}^3.$$

The influence of the vertical stratification of sea water on the gradient component of current velocity is accounted for in the formulas by function $\Phi(z)$, conveniently called the

13. METHOD OF COMPUTING DEEP SEA CURRENTS

stratification function. The expression for $\Phi(z)$ in finite differences has the form

$$\Phi(z) = \frac{\sum_{1}^{n} \overline{\delta_i(z)} \cdot \Delta z_i}{\sum_{1}^{m} \overline{\delta_i(z)} \cdot \Delta z_i}, \qquad (3.42)$$

where $\delta(z) = \rho(0) - \rho(z)$ is the difference in water densities at the "zero" surface and at depth z (the dash on top indicates that the differences are averaged along the vertical between two depths of measurement at a distance of Δz_i), and n and m represent the number of layers between the depths of measurement from the "zero" isobaric surface to the z depth and to the sea surface $z = H$, respectively.

Let us now compute the values of the stratification function $\Phi(z)$ from a given observed vertical distribution of sea water density at hydrological station M. Let us perform our computation for various depths of the "zero" isobaric surface (3000, 2000, 1000, 750, 500, and 400 m), in order to elucidate the dependence of the vertical structure of the current on the location of this surface.

The steps in the computation of function $\Phi(z)$ are shown in Table 10. The second column of the table gives the vertical distribution of density $\sigma_t = (\rho - 1) \cdot 10^3$ (without the correction $\delta\rho$ for the compressibility of sea water). The third column gives the values of $\delta(z) = \rho(0) - \rho(z)$, the fourth, the $\delta(z)$ differences averaged by layers, and the fifth, the products of average $\delta(z)$ differences multiplied by the thickness of the corresponding layer. The sixth column gives the results of the summation of these products from bottom to top, i.e., the sum $\sum_{1}^{n,m} \overline{\delta_i(z)} \Delta z_i$. The final values of the stratification function obtained from the division of values in column six by the sum for the entire layer $0 - H$ are given in the seventh column. The next columns contain $\Phi(z)$ values computed in a similar manner, but with other values for the depth of the "zero" isobaric surface.

The values for coefficients k and a are selected from Tables 7 and 8. The magnitude of argument ε is found in Table 6. In our example $\varepsilon = 0.07$, $a = 7.20 \cdot 10^{-4}$, and

$k = 0.65 \cdot 10^{-5}$. The vertical distribution of current velocity is computed from formulas (3.25) on the assumption that the "zero" surface is located at a depth of 3000 m as shown in Tables 11 and 12. The meridional component of current velocity is computed in Table 11 and the zonal component in Table 12. The vertical distribution of current velocity for other depths of the "zero" surface is computed in a similar manner. The results of computations are summarized in Table 13 and are presented in the form of vector diagrams in Fig. 17.

The computation shows that the pure drift component of current velocity decreases rapidly with depth and

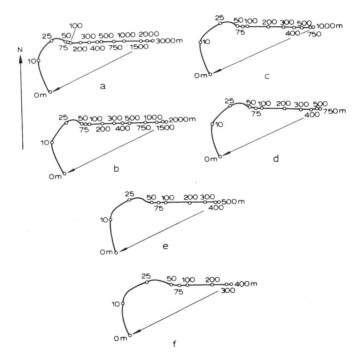

Fig. 17. Vector diagrams of current velocity with various locations of the reference surface:
a — H = 3000 m; b — H = 2000 m; c — H = 1000 m; d — H = 750 m; d — H = 500 m; f — H = 400 m.

changes its direction to the right. It is already practically nonexistent at a depth of the order of 75 m. The gradient component of current velocity retains a constant direction in the entire baroclinic layer; its change with depth is proportional to the stratification function $\Phi(z)$.

To evaluate the accuracy of current velocity computation we must determine how constant the stratification function is in the horizontal direction, because we compute it from the density distribution at one vertical and assume $\Phi(z)$ to be constant within the entire region where the vertical structure of water masses is identical. In the example discussed here, we have the distribution of sea water density along two verticals at points M and N that are 260 miles apart. Vertical density distribution is the same at these stations, i.e., the vertical structure of water masses is identical in this region. The values for the stratification function at point N with $H = 3000$ m are given in Table 14. Comparison of $\Phi(z)$ values computed from data at stations M and N (Tables 10 and 14) shows that deviations do not exceed four units in the second decimal place.

Let us see now what changes in current velocity could result from such horizontal variation of function $\Phi(z)$. According to the results obtained, the gradient component of current velocity is determined from the following formulas:

$$u_g(z) = u_g(H) \cdot \Phi(z);$$
$$v_g(z) = v_g(H) \cdot \Phi(z),$$
(3.43)

where $u_g(H)$ and $v_g(H)$ are the components of the gradient component of current velocity at the sea surface.

As we can see from (3.43), an error in the second decimal place in the determination of the $\Phi(z)$ function introduces an error in velocity that does not exceed 10% of its value. The maximum deviation in the $\Phi(z)$ function determined from data at two hydrological stations is four units of the second decimal place. This yields a maximum error in the determination of the gradient component of current velocity that is 4% of its true value. The error in total current velocity will be even smaller,

Table 10

Computation of function $\Phi(z)$

$H-z$, m	σ_t	$\delta(z)$	3000			Depth of the "zero" surface, m				
			$\overline{\delta(z)}$	$\overline{\delta(z)}\,dz$	l	$\Phi(z)$	2000 $\Phi(z)$	1000 $\Phi(z)$	750 $\Phi(z)$	500 $\Phi(z)$	400 $\Phi(z)$
0	26.04	1.76	1.76	17.6	930.4	1.00	1.00	1.00	1.00	1.00	1.00
10	26.05	1.75	1.75	26.2	912.8	0.98	0.98	0.95	0.95	0.93	0.91
25	26.05	1.75	1.46	36.5	886.5	0.95	0.94	0.88	0.86	0.84	0.79
50	26.64	1.16	1.12	28.0	850.0	0.91	0.87	0.79	0.76	0.71	0.63
75	26.73	1.07	1.08	27.0	822.0	0.88	0.83	0.73	0.69	0.63	0.54
100	26.72	1.08	0.99	99.0	795.0	0.85	0.79	0.67	0.62	0.55	0.46
200	26.90	0.90	0.84	84.0	696.0	0.75	0.66	0.46	0.40	0.30	0.19
300	27.02	0.78	0.72	72.0	612.0	0.66	0.54	0.31	0.23	0.13	0.05
400	27.15	0.65	0.60	60.0	540.0	0.58	0.45	0.19	0.12	0.03	0.00
500	27.26	0.54	0.49	122.5	490.0	0.53	0.37	0.12	0.05	0.00	
750	27.36	0.44	0.41	102.5	367.5	0.40	0.22	0.05	0.00		
1000	27.42	0.38	0.30	150.0	265.0	0.29	0.10	0.00			
1500	27.59	0.21	0.15	75.0	115.0	0.12	0.02				
2000	27.71	0.09	0.06	30.0	40.0	0.04	0.00				
2500	27.77	0.03	0.02	10.0	10.0	0.01					
3000	27.80	0.00			0.0	1.00					

Table 11

Computation of the meridional component of the velocity of a deep current

$H-z$, m	10	25	50	75	100	200	300	400	500	750	1000	1500	2000	2500
$\frac{\partial p}{\partial x} \cdot 10^5$	-11.1	-11.1	-11.1	-11.1										
$\frac{\partial p}{\partial x}$: $\cos a\,(H-z)$	0.76	-0.23	-0.23	0.63										
$10^5 \cos a\,(H-z)$	-8.4	2.5	9.9	-7.0										
$\frac{\partial p}{\partial y} \cdot 10^5$	2.9	2.9	2.9	2.9										
$\frac{\partial p}{\partial y}$: $\sin a\,(H-z)$	0.66	0.97	-0.44	-0.78										
$10^5 \sin a\,(H-z)$	1.9	2.8	-1.3	-2.2										
[3] + [6]	-6.5	5.3	8.6	9.2										
$e^{-a(H-z)}$	0.49	0.17	0.03	0.00										
[7] · [8]	-3.2	0.8	0.2	0.0										
$\Phi(z)$	0.98	0.95	0.91	0.88	0.85	0.75	0.66	0.58	0.53	0.40	0.29	0.12	0.04	0.01
$\frac{\partial p}{\partial x} \cdot 10^5 \Phi(z)$	-10.9	-10.6	-10.1	-9.8	-9.5	-8.3	-7.3	-6.4	-5.8	-4.4	-3.2	-1.4	-0.5	-0.1
[11]−[9]	-7.7	-11.4	-10.3	-9.8	-9.5	-8.3	-7.3	-6.3	-5.8	-4.4	-3.2	-1.4	-0.5	-0.1
$k10^{-5}$	0.65	0.65	0.65	0.65	0.65	3.65	0.65	0.65	0.65	0.65	0.65	0.65	0.65	0.65
[12] · [13]	-5.0	-7.4	-6.7	-6.4	-6.2	-5.4	-4.7	-4.2	-3.8	-2.9	-2.1	-0.9	-0.3	-0.1
$v(H)$	-5.4	-5.4	-5.4	-5.4	-5.4	-5.4	-5.4	-5.4	-5.4	-5.4	-5.4	-5.4	-5.4	-5.4
$v(H) \cdot \Phi(z)$	-5.3	-5.1	-4.9	-4.8	-4.6	-4.0	-3.6	-3.1	-2.8	-2.1	-1.5	-0.7	-0.2	-0.1
$v(z)$	-0.3	2.3	1.8	1.6	1.6	1.4	1.1	1.1	1.0	0.8	0.6	0.2	0.1	0.0

Table

Computation of the zonal compo-

$H-z$, m	10	25	50	75	100	200
$\frac{\partial p}{\partial x} \cdot 10^5$	−11.1	−11.1	−11.1	−11.1		
$\cos a\,(H-z)$	0.66	0.97	−0.44	−0.78		
$\frac{\partial p}{\partial x}\cos a\,(H-z)\cdot 10^5$	−7.3	−10.8	4.9	8.6		
$\frac{\partial p}{\partial y} \cdot 10^5$	2.9	2.9	2.9	2.9		
$\sin a\,(H-z)$	0.76	−0.23	−0.90	0.63		
$\frac{\partial p}{\partial y}\sin a\,(H-z)\cdot 10^5$	2.2	−0.7	−2.6	1.8		
[3] + [6]	−9.5	−10.1	7.5	6.8		
$e^{-a(H-z)}$	0.49	0.17	0.03	0.00		
[8] · [7]	4.6	−1.7	0.2	0.0		
$\Phi(z)$	0.98	0.95	0.91	0.88	0.85	0.75
$\frac{\partial p}{\partial x}\Phi(z)\cdot 10^5$	2.8	2.8	2.7	2.6	2.5	2.2
[11] −[9]	−1.8	1.1	2.9	2.6	2.5	2.2
$k \cdot 10^{-5}$	0.65	0.65	0.65	0.65	0.65	0.65
[13] · [12]	−1.2	0.7	1.9	1.7	1.6	1.4
$u(H)$	−15.4	−15.4	−15.4	−15.4	−15.4	−15.4
$u(H)\cdot \Phi(z)$	−15.1	−14.7	−14.1	−13.6	−13.2	−11.5
$u(z)$	−16.3	−14.0	−12.2	−11.9	−11.6	−10.1

because the pure drift component of current velocity does not depend on the structure of the density field.

Comparison of vertical distributions of current velocity computed for various depths of the "zero" isobaric surface shows that the result of the computation depends largely on how accurately the depth of the "zero" surface has been determined. Consequently, in order to use the foregoing method of computing deep currents from the surface current, we must be able to estimate, even though roughly, the depth of penetration of wind-driven currents (depth of the "zero" surface). The next paragraph is devoted to the solution of this problem.

It is not necessary to know the absolute value of current velocity at the sea surface to compute the velocities of deep currents. The computations can be performed if we know the difference between current velocities at any

13. METHOD OF COMPUTING DEEP SEA CURRENTS

nent of velocity of a deep current

300	400	500	750	1000	1500	2000	2500
0.66	0.58	0.53	0.40	0.29	0.12	0.04	0.01
1.9	1.7	1.5	1.1	0.8	0.4	0.1	0.0
1.9	1.7	1.5	1.1	0.8	0.4	0.1	0.0
0.65	0.65	0.65	0.65	0.65	0.65	0.65	0.65
1.2	1.1	0.9	0.7	0.5	0.3	0.1	0.0
−15.4	−15.4	−15.4	−15.4	−15.4	−15.4	−15.4	−15.4
−10.2	− 8.8	− 8.1	− 6.1	− 4.4	− 1.9	− 0.7	− 0.2
− 9.0	− 7.7	− 7.2	− 5.4	− 3.9	− 1.6	− 0.6	− 0.1

two depths. Let us assume that the difference in current velocities at the sea surface and some depth z_1 is known from observation. Then, subtracting $u(H)$ and $v(H)$ from both parts of formula (3.25), respectively, and resolving the result relative to the components of current velocity at the sea surface, we obtain

$$u(H) = \frac{[u(H) - u(z_1)] + k\Phi(z_1)\frac{\partial p}{\partial y}}{1 - \Phi(z_1)} +$$

$$\frac{ke^{-a(H-z_1)}\left[\frac{\partial p}{\partial x}\sin a(H-z_1) - \frac{\partial p}{\partial y}\cos a(H-z_1)\right]}{1 - \Phi(z_1)};$$

$$v(H) = \frac{[v(H) - v(z_1)] - k\Phi(z_1)\frac{\partial p}{\partial x}}{1 - \Phi(z_1)} - \qquad (3.44)$$

$$\frac{ke^{-a(H-z_1)}\left[\frac{\partial p}{\partial x}\cos a\,(H-z_1)+\frac{\partial p}{\partial y}\sin a\,(H-z_1)\right]}{1-\Phi(z_1)}.\qquad\begin{array}{c}(3.44)\\ \text{(cont'd)}\end{array}$$

Equations (3.44) make it possible to compute the velocity of the surface current from the measured difference in the velocities of the current at the sea surface and at depth $(H-z_1)$. Substituting expression (3.44) in (3.25), we obtain formulas for computing the vertical distribution of current velocity within the baroclinic layer from the measured difference in current velocities at two depths.

Numerous data are now available for current velocity at two depths measured simultaneously on a drifting ship by the Makarov-Nansen method. Using this material we can compute not only the absolute velocity of a current at the sea surface, but also the vertical distribution of current velocity to considerable depths.

Indeed, the current velocity $\vec{w}_c$ measured from a drifting ship is the sum of the aperiodic current $\vec{w}$, the tidal current $\vec{w}_{n-0}$, and the so-called "wake" current $\vec{w}_k$ (a value that is opposite in direction to the ship's drift velocity vector). We can assume with a sufficient degree of accuracy that the velocity of a tidal current over great depths in the open sea does not depend on the vertical coordinate. In this case the tidal component of current velocity and the velocity of the "wake" current must cancel each other when the difference between current velocities is computed

$$\vec{w}_c(H)-\vec{w}_c(z_1)=\vec{w}(H)-\vec{w}(z_1),\qquad(3.45)$$

since

$$\vec{w}_{n-0}(H)=\vec{w}_{n-0}(z_1);$$
$$\vec{w}_k(H)=\vec{w}_k(z_1).$$

It is very important that current velocities at two depths be measured simultaneously or within a time

Table 13

Computation of the horizontal components of current velocity, cm/sec

Depth of the "zero" surface, м

$H-z$, m	3000		2000		1000		750		500		400	
	u	v	u	v	u	v	u	v	u	v	u	v
0	−15.4	−5.4	−15.4	−5.4	−15.4	−5.4	−15.4	−5.4	−15.4	−5.4	−15.4	−5.4
10	−16.3	−0.3	−16.4	−0.4	−16.1	−0.4	−16.0	−0.4	−15.7	−0.4	−15.6	−0.4
25	−14.0	2.3	−14.0	2.2	−13.2	2.1	−13.0	2.1	−12.6	2.1	−11.9	2.0
50	−12.2	1.8	−11.8	1.7	−10.8	1.5	−10.3	1.5	−9.6	1.4	−8.5	1.3
75	−11.9	1.6	−11.4	1.5	−10.0	1.3	−9.5	1.3	−8.6	1.1	−7.4	1.0
100	−11.6	1.6	−10.9	1.4	−9.3	1.2	−8.6	1.1	−7.6	1.0	−6.1	0.8
200	−10.1	1.4	−9.0	1.2	−6.4	1.1	−5.4	0.8	−4.1	0.6	−2.6	0.4
300	−9.0	1.1	−7.4	1.0	−4.2	0.5	−3.1	0.4	−1.8	0.3	−0.6	0.0
400	−7.7	1.1	−6.1	0.8	−2.6	0.4	−1.6	0.2	−0.4	0.1	0.0	0.0
500	−7.2	1.0	−5.1	0.7	−1.6	0.2	−0.7	0.1	0.0	0.0	—	—
750	−5.4	0.8	−3.0	0.4	−0.3	0.1	−0.0	—	—	—	—	—
1000	−3.9	0.6	−1.4	0.2	−0.0	0.0	—	—	—	—	—	—
1500	−1.6	0.2	−0.3	0.1	—	—	—	—	—	—	—	—
2000	−0.6	0.1	−0.0	0.0	—	—	—	—	—	—	—	—
2500	−0.1	0.0	—	—	—	—	—	—	—	—	—	—
3000	0.0	0.0	—	—	—	—	—	—	—	—	—	—

interval during which the change in the ship's drift velocity and in the tidal current can be neglected.

Table 14

Values of function $\Phi(z)$

$H-z$, m	0	10	25	50	75	100	200	300
$\Phi(z)$	1.00	0.98	0.95	0.91	0.87	0.85	0.74	0.66
$H-z$, m	400	500	750	1000	1500	2000	2500	3000
$\Phi(z)$	0.59	0.53	0.40	0.30	0.18	0.08	0.02	0.00

In describing the method of computing deep sea currents from surface currents, we emphasized the applicability of this method only to the computation of steady currents or currents that are very stable in respect to time. Now let us use individual measurements for the computation. The use of individual measurements instead of measurements averaged over a long period of time is not strictly justified. However, theoretical investigations show that it takes several hours of steady wind action to produce a pure drift component of current velocity in the entire layer of friction, whereas it requires approximately a month to produce the gradient component of current velocity in the entire baroclinic layer (Shtokman and Tsikunov, 1954). This is probably the explanation for the fact that rather steady water circulation systems form in oceans and seas though the wind fields over them are very unstable. Observations show that the dynamic topography of isobaric surfaces retains its general features for a very long time (Post, 1954).

Proceeding from this, we found it possible to use individual but simultaneous measurements of current velocity at two depths for the computation of surface and deep currents, provided the wind was steady in direction and force 5-6 hours before the measurements were made. Better results can be expected in regions with strong gradient currents.

14. INVESTIGATION OF THE DEPTH OF PENETRATION OF A WIND-DRIVEN OCEAN CURRENT

Let us now try to estimate the depth of penetration of wind circulation in the sea on the basis of the above method of computing deep currents from surface currents. To do this, let us integrate formulas (3.25) with respect to z from the lower boundary of the baroclinic layer ($z = 0$) to the free surface of the sea ($z = H$):

$$u(H) + k\frac{\partial p}{\partial y} = \left(S_x - \frac{k}{2a}\left\{\frac{\partial p}{\partial x}[1 - e^{-aH}(\sin aH + \cos aH)] - \frac{\partial p}{\partial y}[1 + e^{-aH}(\sin aH - \cos aH)]\right\}\right)F(H),$$

$$v(H) - k\frac{\partial p}{\partial x} = \left(S_y - \frac{k}{2a}\left\{\frac{\partial p}{\partial x}[1 - e^{-aH}(\cos aH - \sin aH)] + \frac{\partial p}{\partial y}[1 - e^{-aH}(\sin aH + \cos aH)]\right\}\right)F(H),$$

(3.46)

where

$$F(H) = \frac{\int_0^H \delta(z)\,dz}{\int_0^H dz \int_0^z \delta(z)\,dz}; \quad (3.47)$$

S_x and S_y are the components of the full current along the coordinate axes X and Y.

The second terms in the left and right parts of expressions (3.46) define the pure drift components of the velocity of the surface current and of the full current, respectively. Having estimated the order of magnitude of the values entering expressions (3.46), we find that the pure drift component of the full current is smaller by at least an order of magnitude than the values of the total full current. Neglecting the small terms, we obtain the final formulas for computing the velocity components of the surface current from the given values of the full current, the atmospheric pressure gradient, and the function describing the vertical structure of the sea water density field:

$$u(H) = S_x F(H) - k \frac{\partial p}{\partial y};$$

$$v(H) = S_y F(H) + k \frac{\partial p}{\partial x}.$$

(3.48)

Since we used a density model in the derivation, it is sufficient to know the observed vertical density distribution at only one point to account for the effect of the vertical stratification of sea water. We must also assume that the stratification function $F(H)$ computed for this water column is constant along the horizontal in the region where the vertical structure of the water masses is identical. The horizontal change of density due to the adjustment of the density field to the current system is accounted for explicitly in formulas (3.48) by the value of the full current, because the latter depends both on the dimensions and configuration of the basin and on the wind field inducing currents.

To determine the depth of penetration of the current we must compute from formulas (3.48) a number of values for the velocity of the surface current with various assumed depths of the lower boundary of the current and compare this current velocity with the actually observed current velocity at the sea surface. Such a comparison uniquely determines H. We can do the same by determining the stratification function $F(H)$ by (3.48) from the measured current velocity at the sea surface, the full current, and the atmospheric pressure gradient, and compare this function with the stratification function computed from the observed distribution of sea water density. But the depth of penetration thus found will be reliable only if the computed full current and the measured current velocity describe the same dynamic state of the water mass, i.e., if they correspond.

We shall examine the method of determining the depth of penetration of a current in more detail by an actual example. Let us take the North Atlantic with synoptic and hydrological conditions for May as such an example. The region under study is represented in Fig. 18, where

14. INVESTIGATION OF THE DEPTH OF PENETRATION 103

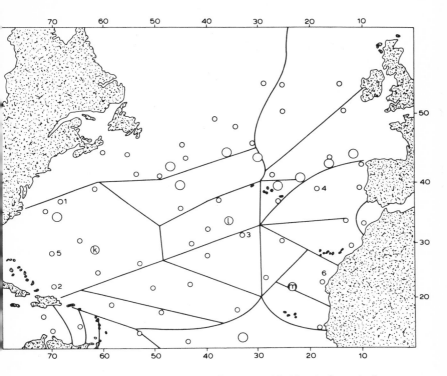

Fig. 18. North Atlantic Ocean. Regions with identical vertical structure of water masses (according to Jacobsen): k, l, m—stations: the numbers in the circles correspond to the curves in Fig. 24.

the small circles indicate points with a long-period average (for May) surface current velocity that is known from observation. We shall determine the position of the lower boundary of the current for these points.

To account for the effect of the vertical stratification of sea water, we must divide the area into regions that have an identical vertical water mass structure and then compute the stratification function for each region. The North Atlantic has been divided into regions using T-S curves by Jacobsen (1929). Figure 18 gives the boundaries of such regions, according to Jacobsen. The large circles denote hydrological stations whose data were used to compute the stratification function. As we can see from Fig. 18, the hydrological stations are unevenly distributed in the region under consideration. The vertical structure of density remains unknown for approximately half the region. We shall demonstrate below that the stratification function varies little in the horizontal direction, and therefore, for our purpose, there is no need for precise regional breakdown.

An example of the computation of the stratification function $F(H)$ is shown in Table 15, where $F(H)$ is computed for hydrological station k. It is assumed that the lower boundary of the current is at 2000 m. The vertical distribution of density σ_t (without correction δ_p for the compressibility of sea water), given in the second column of

Table 15

Computation of function F (H)

$H-z$, m	σ_t	$\delta(z)$	$\overline{\delta(z)}$	$\overline{\delta(z)}dz$	I	I_{av}	$I_{av}dz$
1	2	3	4	5	6	7	8
0	25.60	2.21			1348		
25	25.67	2.14	2.18	54	1294	1321	33,025
50	25.95	1.86	2.00	50	1244	1269	31,725
75	26.16	1.65	1.76	44	1200	1222	30,550
100	26.20	1.61	1.63	41	1159	1180	29,500
125	26.23	1.58	1.60	40	1119	1139	28,745
150	26.24	1.57	1.58	39	1080	1119	27,500
200	26.26	1.55	1.56	78	1002	1041	52,050
300	26.33	1.48	1.51	151	851	926	92,600
400	26.42	1.39	1.43	143	708	780	78,000
500	26.57	1.24	1.32	132	576	642	64,200
600	26.63	1.18	1.21	121	455	516	51,600
700	26.75	1.06	1.12	112	342	399	39,900
800	26.91	0.90	0.98	98	245	294	29,400
900	27.12	0.69	0.80	80	165	205	20,500
1000	27.29	0.52	0.60	60	105	135	13,500
1200	27.62	0.19	0.35	70	35	70	14,000
1500	27.79	0.02	0.10	30	5	20	6,000
2000	27.81	0.00	0.01	5	0	2.5	7,500
							650,025

$F(H) = 2.078 \cdot 10^{-5}$

14. INVESTIGATION OF THE DEPTH OF PENETRATION

the table, is used as the initial vertical. The third column gives $\delta(z)$ values, the fourth, $\delta(z)$ differences averaged by layers, and the fifth, the products of average $\delta(z)$ values multiplied by the thickness of the corresponding layer. The sixth column gives the summation of values in the fifth column, or $\int^{z} \delta(z)dz$ values with $0 \leqslant z \leqslant H$. The seventh and eighth columns contain, respectively, the values of the sixth column averaged by layers and their products multiplied by the thickness of the layer between two successive depth levels. The value of the stratification function is obtained by dividing the number in the first line of the sixth column by the sum of the values in the eighth column. To convert the result to the cgs system, we must multiply it by 10^{-2}.

The values of the stratification function for other assumed depths of the lower current boundary were computed in a similar manner. As a result, a tabular representation of function $F(H)$ was obtained. This function is represented graphically by a monotonic curve asymptotically approaching the ordinate axis and a straight line segment parallel to the x-axis (Fig. 19). The graph shows that the stratification function changes rapidly with small H values (in the surface layer where the vertical density gradient is steep) and is practically independent of H at $H > 1500$ m (where the vertical density gradient is extremely small).

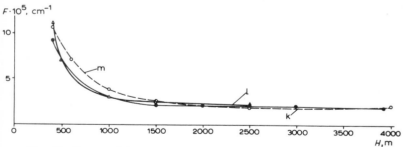

Fig. 19. Curves of the stratification function F(H) at points k, l, m.

Comparison of $F(H)$ values computed from the vertical distribution of sea water density at various points in the ocean shows that the stratification function depends little on position. The curves plotted on a single graph show a common variation pattern and agree rather well in magnitude (Fig. 19 shows curves for three stations located in different parts of the ocean). They agree so well at relatively close points that deviations, in all probability, do not exceed the limits of accuracy of the computation. Hence we may conclude that the stratification function describes only the most general features of the vertical structure of the water density field without giving its details.

The derivatives of atmospheric pressure along the horizontal coordinates $\frac{\partial p}{\partial x}$ and $\frac{\partial p}{\partial y}$ and the S_x and S_y components of the full current were determined by the method of finite differences at those points at which the depth of penetration of the wind-driven current was estimated. To do this we used the long-term mean atmospheric pressure map for May and the map of the function of the full current (Sarkisyan, 1954) that are presented in Figs. 20 and 21, respectively. The distribution of surface current velocity is shown in Fig. 22. The numbers above the arrows represent current velocity in miles per day.

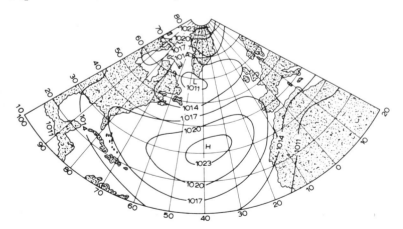

Fig. 20. Map of long-term mean atmospheric pressure for May.

To determine the depth of penetration of the current for each point from formulas (3.48), we computed the

14. INVESTIGATION OF THE DEPTH OF PENETRATION 107

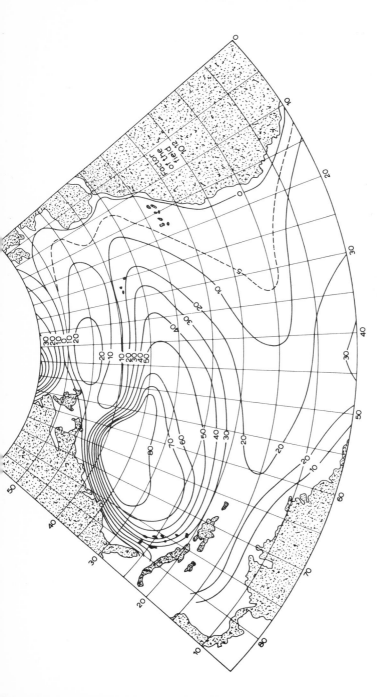

Fig. 21. Chart of full steady currents in the North Atlantic in May (according to A. S. Sarkisyan).

108   III. INVESTIGATION OF THE VERTICAL VELOCITY

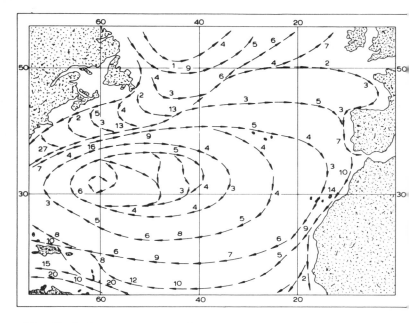

Fig. 22. Surface currents in the North Atlantic (May). The numbers above the arrows indicate current velocity in miles per day.

values of the stratification function $F(H)$. Then, according to the regional breakdown of the area under study, we took the values of the depths of penetration H of the current from the $F(H)$ curves as a function of H (Fig. 19). We used the nearest or several of the nearest stations for points located in areas without a single hydrological station. The simultaneous use of several hydrological stations did not, as a rule, lead to significant discrepancies in the determination of H.

The results of computations were used to construct a relief map of the lower boundary of the wind-induced current (Fig. 23). The numbers on the contours indicate the depth of the lower boundary of the current in meters.

Comparison of the relief map with the map of the function of the full current indicates that the lower boundary of the current rises in regions where the full current is sharply defined. In the central part, where the

14. INVESTIGATION OF THE DEPTH OF PENETRATION

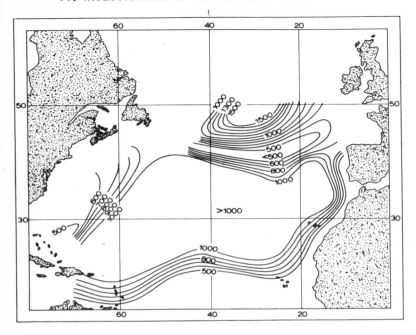

Fig. 23. Relief of the lower boundary of the baroclinic layer in the North Atlantic.

full current is weak, the lower boundary of the current is deeper and forms a bowl-shaped depression. In the south the thickness of the layer embraced by the current decreases toward the equator. This was noted earlier by Defant (1941b). In our opinion, it is too early to draw any conclusions from the results obtained because we are not certain whether the velocities of the surface current and the magnitudes of the full current, used for computation, fully correspond. The map of the functions of the full current is too schematic, and the distribution of surface current velocities hardly describes the field of steady currents at the sea surface in sufficient detail. For these reasons it is difficult to estimate how well the results obtained approximate the actual location of the lower boundary of the current in the North Atlantic. Our example is only an illustration.

Let us now examine the practical conclusions that we can draw from the actual material used in the example. According to (3.48), the absolute value of the gradient component of current velocity at the sea surface, $w_g(H)$, is given by the formula

$$w_g(H) = SF(H), \qquad (3.49)$$

where S is the absolute magnitude of the full current. The same component of current velocity at depth z is

$$w_g(z) = SF(H)\Phi(z) \qquad (3.50)$$

[a similar expression was obtained earlier by V. B. Shtokman (1951)].

The last formula shows that the vertical variation of current velocity is directly associated with the vertical structure of the water density field. Current velocity changes rapidly in the upper layers immediately adjoining the surface. At greater depths, where the vertical density gradient is small, current velocity cannot change significantly in the vertical direction. Thus we may conclude that the horizontal pressure gradient caused by the dynamic slope of the free surface of the sea is almost fully compensated by the adjustment of the field of mass in the water layer with a steep vertical density gradient. In the opposite case we would have to assume that the wind-driven current penetrates to greater ocean depths, because the horizontal pressure gradient and, with it, the gradient component of current velocity cannot vary significantly in the vertical direction in weakly stratified deep water.

Thus a wind-driven current is practically confined to the layer with a strong density gradient, provided, of course, that the water is sufficiently stratified along the vertical and is not nearly homogeneous from the surface to the bottom. The lower boundary of the current obviously cannot be deeper than the surface or the layer that divides regions with strong and weak vertical density gradients. But sea water density varies rather uniformly with depth. To elucidate the concepts of "strong" and "weak" density gradients let us examine Fig. 24, which is a graphical

14. INVESTIGATION OF THE DEPTH OF PENETRATION

representation of the variation of the gradient component of current velocity at the sea surface as a function of the depth of the lower boundary of the current. The numbers on the curves are the serial numbers of stations to which these curves refer (Fig. 18). The depths of the lower boundary of the current are plotted along the x-axis and current velocity, in cm/sec, along the y-axis. Figure 24 shows that starting from a depth of 1500-2000 m, a further increase in the thickness of the layer embraced by the current does not lead to a significant change in the velocity of the surface current. Current velocity as a function of H is represented in the same manner at any other level. Thus, in our example the vertical density gradient must be assumed to be small, starting from a depth of 1500-2000 m. This depth corresponds to the value of H in the region where the curves in Figs. 19 and 24 straighten out. Current velocity must be negligibly small at depths below 2000 m. If this is not so and if current velocity below 2000 m is still high, the current must inevitably reach bottom with a high velocity, because weakly stratified deep water is incapable of noticeably modifying velocity.

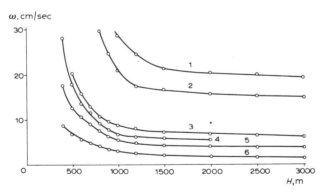

Fig. 24. Dependence of current velocity at the surface on the depth of the lower boundary of the current.

We noted earlier that the depth of penetration of a wind-driven current (the lower boundary of the baroclinic layer) can be identified with the surface at which current velocity constitutes some given portion of current velocity at the sea surface. The relationship between gradient components of current velocity at depth z and at the sea surface can be obtained from expressions (3.39) and (3.50), hence

$$\frac{w_g(z)}{w_g(H)} = \frac{\int_0^z \delta(z)\,dz}{\int_0^H \delta(z)\,dz}. \qquad (3.51)$$

Formula (3.51) includes two unknowns: the depth of the surface $z = z(x, y)$ at which current velocity constitutes a given fraction of current velocity at the sea surface, and the depth of penetration of the current, H. For this reason H can be computed only if z has a known value or number of values.

Proceeding from the fact that, strictly speaking, a wind-driven current can penetrate to any depth, we must expect that z, corresponding to the depth of the surface where $\frac{w_g(z)}{w_g(H)}=$ const, must indefinitely approach its limiting value with increasing H. The results of computations confirm this assumption.

Table 16 gives the depth of $\frac{w_g(z)}{w_g(H)} = 0.2$ and 0.1 surfaces with a number of given H values as computed from vertical density distribution. As we can see from the table, the depth where current velocity constitutes the given portion of the velocity of the surface current tends toward its limiting value with increasing H. This limit can always be determined if observations at hydrological stations are made to great depths.

Table 16

$\frac{w_g(z)}{w_g(H)}$	H, m			
	2500	3000	3500	4000
0.2	800	870	900	900
0.1	1020	1100	1170	1200

The practical application of the results obtained in this paragraph will be demonstrated in Chapter 6 when we construct a dynamic chart of the sea surface in the region of the Kurile-Kamchatka basin in the Pacific. It should be noted here that this characteristic of the vertical current velocity distribution was obtained on the assumption that the velocities of the steady gradient current are insignificant at the bottom of the sea. The density field as such cannot indicate the position of the surface at which the velocity of the gradient current is zero, because the density field can be used only to judge the variation of current velocity with depth.

The method described in this paragraph can be used to determine the optimum depth of the "zero" surface for computation of currents by the dynamic method. By locating the reference surface below this depth, we will only increase computation errors and approximation to the absolute magnitude of current velocity will not be achieved, because deep water has little effect on the horizontal pressure gradient.

15. METHOD OF INCREASING THE ACCURACY OF DYNAMIC COMPUTATION

A way of improving the accuracy of sea current computations by the dynamic method, where the depth of the layer of no motion is determined reliably but is located at an exceedingly great depth, follows directly from the foregoing. In the preceding chapter we demonstrated that placement of the reference surface at a great depth leads to large computational errors, since the current velocity error is directly proportional to the depth H of the reference surface. If the "zero" surface is at a depth of several thousands of meters, computational errors may completely distort the result.

The difference between current velocity at the sea surface $z = H$ and at the z depth is given in the dynamic method by the expression:

$$v(H) - v(z) = \Delta D \cdot M, \qquad (3.52)$$

where ΔD is the difference in dynamic heights at two neighboring hydrological stations, i.e., the difference in distances $(H-z)$ at neighboring stations as expressed in dynamic units; $M = (2\omega L \sin \varphi)^{-1}$. The value for z is selected arbitrarily.

Expressions (3.51) and (3.52) form a closed system of equations, hence

$$v(H) = \Delta D \cdot M \cdot \frac{\int_0^H \delta(z)\,dz}{\int_z^H \delta(z)\,dz}; \qquad (3.53)$$

$$v(z) = \Delta D \cdot M \cdot \frac{\int_0^z \delta(z)\,dz}{\int_z^H \delta(z)\,dz}. \qquad (3.54)$$

The relative mistake in the computation of current velocity at the sea surface by formula (3.53) is estimated from

$$\frac{dv(H)}{v(H)} = \frac{2(H-z)}{\Delta D}\,dv_t + \frac{2d\sigma_t}{\int_0^H \delta(z)\,dz} \cdot \frac{\left[z\int_0^H \delta(z)\,dz - H\int_0^z \delta(z)\,dz\right]}{\int_0^H \delta(z)\,dz}. \qquad (3.55)$$

The first term of formula (3.55) is the relative mistake in the computation of current velocity at the sea surface relative to the current velocity at the z depth. Its magnitude depends largely on the selection of the value for the $(H-z)$ difference. Since z is selected arbitrarily, the first term in (3.55) can be made relatively small by assigning small values to $(H-z)$. The magnitude of the second term of formula (3.55) depends little on depth H, i.e., the depth of the "zero" surface. It is sufficient to say that when the lower boundary of the current is placed at a depth of 5000 m, the magnitude of this term does not exceed 15-20% and it decreases with decreasing depth of the "zero" isobaric surface.

15. INCREASING ACCURACY OF DYNAMIC COMPUTATION   115

Table 17 gives values, expressed in percentages, of the second term of formula (3.55) with various H and z. The vertical distribution of density σ_t, used to compute the table, is given in the second column.

Table 17

Magnitude of the second term of formula (3.55), %

$H-z$, m	σ_t	Depth of the "zero" surface H, m						
		5000	4000	3000	2500	2000	1500	1000
0	26.51	0.0	0.0	0.0	0.0	0.0	0.0	0.0
10	26.54	0.2	0.3	0.1	0.2	0.2	0.2	0.0
25	26.55	0.6	0.6	0.4	0.5	0.4	0.5	0.4
35	26.56	0.8	0.9	0.6	0.6	0.7	0.6	0.4
50	26.60	1.1	1.1	0.9	0.9	0.9	0.8	0.6
75	26.66	1.7	1.7	1.2	1.2	1.3	1.3	0.8
100	26.72	2.1	2.2	1.6	1.6	1.8	1.6	1.0
125	26.76	2.6	2.7	2.0	1.9	2.0	2.1	1.3
150	26.82	3.0	3.1	2.4	2.3	2.4	2.2	1.7
175	26.88	3.4	3.5	2.7	2.6	2.6	2.6	1.7
200	26.92	3.8	4.0	3.0	2.9	3.0	2.9	1.9
250	27.01	4.5	4.6	3.5	3.4	3.4	3.2	2.1
300	27.06	5.1	5.2	3.9	3.7	3.8	3.5	2.1
400	27.15	6.3	6.3	4.6	4.4	4.4	4.0	2.3
500	27.24	7.0	7.2	5.2	5.0	4.8	4.2	2.3
600	27.29	7.7	7.9	5.6	5.3	5.1	4.3	1.9
750	27.41	9.5	8.6	6.1	5.6	5.2	4.0	1.3
1000	27.49	10.3	9.3	6.3	5.6	4.8	3.2	0.0
1500	27.63	10.8	9.4	5.8	5.3	3.0	0.0	—
2000	27.69	10.2	8.4	4.3	2.6	0.0	—	—
2500	27.74	9.1	6.8	2.3	0.0	—	—	—
3000	27.77	7.5	4.7	0.0	—	—	—	—
4000	27.78	3.9	0.0	—	—	—	—	—
5000	27.80	0.0	—	—	—	—	—	—

Thus, when the reference surface is located at a depth of several thousand meters, the gradient component of current velocity at the sea surface is computed from formula (3.55) with an error that does not exceed 50% of the velocity value. In ordinary dynamic computation the mistake in current velocity under the same conditions is several times the actual current velocity. It is quite clear that not only the sea surface, but also any other isobaric surface at depth, can be considered as the H level.

In using this method to compute gradient currents, we should not forget that it is based on Shtokman's density model. This means that it is suitable only for single-layer circulation. Better results can be expected when the requirements of the law of the parallelism of solenoids are satisfied in the region under study, because then the structure of the density field agrees most closely with the model.

CONCLUSIONS

1. The vertical variation of the velocity of a gradient current depends on sea water stratification. In the layers underlying the surface, where the vertical density gradient is strong, velocity attenuates considerably with depth. At great depth, where the water is almost homogeneous, the velocity of the gradient current does not change noticeably along the vertical.

2. The depth of the reference surface that is optimum for the dynamic method, i.e., the depth below which the water density field has no significant effect on the gradient component of current velocity, can be estimated from the form of the stratification function $F(H)$. This depth corresponds to the H value in the region where the curve of the stratification function $F(H)$ straightens out.

3. Function $\Phi(z)$, which describes the relationship of the moduli of the gradient component of current velocity at two depths, changes little in the horizontal direction in regions where the vertical structure of water masses is identical. Consequently, when currents are computed by the dynamic method relative to a horizontal reference surface, current velocities are obtained with an accuracy to the constant multiplier.

4. The method of increasing the accuracy of dynamic computation can be used when currents are computed in deep-water regions where the "zero" surface is at a great depth and the water density field corresponds to the model.

CHAPTER IV

Methods for Computing the "Zero" Surface in the Sea

16. GENERAL EVALUATION OF METHODS FOR DETERMINING THE LAYER OF NO MOTION IN THE SEA

The structure of the current velocity field in a sea depends on the forces and factors that induce currents or affect the vertical distribution of current velocity. The wind velocity over the sea, sea water stratification, the coriolis force, the influence of shores and bottom relief, etc., differ in various parts of the sea. Being the geometrical location of sea water particles that possess small or zero velocity, the "zero" surface cannot be horizontal over a great distance. It has a complex form which depends on the distribution of forces and factors affecting water particle motion.

This point of view is accepted by all oceanographers, but a horizontal reference surface is used, as a rule, in dynamic computations because of the lack of a reliable method of determining the layer of no motion. Sometimes the lower level of measurement is simply used for this purpose. Sometimes the water density field is analyzed first and the horizontal reference surface is placed in the layer where density is least disturbed (Soule, 1939). Under these conditions absolute computation of current velocities by the dynamic method is obviously out of the question, since current velocity can differ from zero at the reference surface.

To obtain absolute results from current computations by the dynamic method, we must always begin from a strictly justified selection of the layer of no motion in the sea or from the known distribution of current velocity along some surface. The latter would be more reliable, but direct measurements of current velocity at sea involve great difficulties. Rapid surveys of currents from one ship over a large area in tidal seas are not possible even with modern equipment. Moreover, the current at a given moment is characterized by a directly measured velocity, whereas the dynamic method deals with the gradient component of a steady current. There are only a few examples known in oceanography where dynamic analysis of measurements was applied to reduce the current computed by the dynamic method to absolute values, using measured current velocities (Wüst, 1924; Swallow and Worthington, 1959).

In the last three decades a number of scientists have developed methods for determining the "zero surface" (layer of no motion). An indirect approach to the solution is common to most of these methods; the authors tried to find a characteristic that would point to absent or minimum motion in some water layer. We shall discuss the main methods of determining the layer of no motion below, and touch indirectly on the other methods.

17. IDENTIFICATION OF THE LAYER OF NO MOTION IN THE SEA WITH THE INTERMEDIATE OXYGEN MINIMUM (DIETRICH'S METHOD)

In many regions of the World Ocean there is a layer at intermediate depths with a minimum content of dissolved oxygen. In the Atlantic Ocean this layer can be traced for a long distance, from 45°S lat. to 55°N lat. The depth of this layer varies in different parts of the ocean, but is confined to depths ranging from 100 m to 950 m.

There is still no generally accepted explanation for the formation of the intermediate oxygen minimum in the ocean, even though many investigations have been devoted to the solution of this problem (Jacobsen, 1916; Wattenberg, 1929; Seiwell, 1934; Danois, 1934; Wüst, 1935). From these studies it follows that the oxygen minimum layer at intermediate depths results either from the relatively weak motion of water there (boundary between

17. IDENTIFICATION OF THE LAYER OF NO MOTION

two differently directed systems of circulation), or from a combination of biological and physical factors leading to an irregular distribution of oxygen-consuming organic material in the water.

Jacobsen (1916) was the first to advance the idea that the oxygen minimum in sea water corresponds to the layer with minimum horizontal water motion. It was further developed and applied in practice by Wüst (1935) and Dietrich (1936). These authors believe that, with the exception of the relatively thin upper layer, oxygen is lost from sea water primarily as a result of the oxidation of organic matter. Moreover, they believe that the distribution of organic matter is uniform with depth and, consequently, oxygen consumption on oxidation must be uniform and independent of depth. The irregularity in the vertical distribution of oxygen in sea water must then be entirely due to dynamic causes. The layer of the oxygen minimum must correspond to the layer with the minimum supply of oxygen. But since oxygen in sea water is redistributed by currents and turbulent processes induced by them, it follows that the oxygen minimum must correspond to the layer of minimum motion in the sea.

This concept was used by Wüst (1935) to distinguish water masses in the Atlantic Ocean. On the same basis Dietrich (1936) computed currents in the Gulf Stream and in the Agulhas current by the dynamic method. He assumed that the "zero" surface corresponded to the oxygen minimum layer. The relief of individual isobaric surfaces along the Chesapeake Bay-Bermuda profile computed by Dietrich is presented in Fig. 25, which includes for comparison the isolines of isobaric surfaces with the reference surface at a depth of 1000 and 2000 decibars. The path of the Gulf Stream is located between stations 1227 and 1229. Dynamic computation using the oxygen minimum as the absolute reference surface yields a countercurrent with a velocity of 28 cm/sec at a depth of 1000 m. At station 1229 the water column below 1000 m is heavier than at station 1227; consequently the countercurrent reaches the near-bottom depths here, and its velocity increases toward the bottom. According to Dietrich's computations, the water transport in the Gulf Stream from the sea surface to the oxygen minimum layer is $31 \cdot 10^6 \text{ m}^3/\text{sec}$, whereas the water transport in the countercurrent below the Gulf Stream is twice that ($78 \cdot 10^{-6} \text{ m}^3/\text{sec}$), which is difficult to believe.

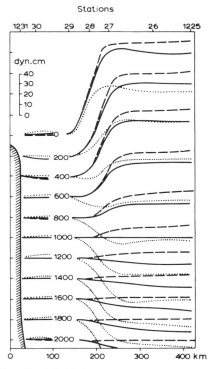

Fig. 25. Relief of isobars computed relative to various reference surfaces along the profile from the Chesapeake Bay to Bermuda: – – – – relative to 2000 decibars; ——— 1000 decibars; . . . relative to the oxygen minimum.

Iselin (1936) computed current velocities by the dynamic method in the same region and from the same data, using a reference surface at 2000 m. His results show that in the region of the intermediate oxygen minimum the gradient current velocity is high and almost constant. Since the vertical velocity gradient is independent of the

17. IDENTIFICATION OF THE LAYER OF NO MOTION

selection of the reference surface, the dynamic activity of the layer of the oxygen minimum cannot be small. Even if current velocity in the oxygen minimum were insignificant, the presence of a strong vertical current velocity gradient would lead to intense turbulent mixing and to the destruction of the intermediate oxygen minimum. But since this does not take place, advection must inevitably occur in this layer, i.e., a noticeable current must exist.

Moreover, the assumptions that are objectionable are that the vertical distribution of oxygen-consuming organic matter is uniform and that oxygen consumption is independent of the oxygen content in the water. Sverdrup (1938) showed that oxygen consumption in a steady state cannot be independent of depth. If the oxygen supply is minimal at some depth, then oxygen consumption at this depth must also be minimal. The distribution of oxidizing organic matter also cannot be independent of the vertical coordinate. We know from observation that detritus accumulates in very stable layers.

Seiwell's studies (1937) are of great importance to the refutation of the dynamic explanation of the intermediate oxygen minimum in sea water. He examines the causes for minimum oxygen concentrations at intermediate depths in the northwestern part of the Atlantic Ocean. Analysis of a large number of observations shows that the minimum is located along the isopycnal surface $\sigma_t = 27.23 \pm 0.08$, which is located in an interval of great vertical stability. This leads to the idea that organic matter accumulates in layers with strong vertical gradients and that oxygen consumption is intensified there. This results in an oxygen minimum at these depths. It would be more correct to say that the oxygen minimum in sea water is the result of the ratio between oxygen supply and consumption. Seiwell examines a hypothetical example that shows that even when oxygen supply and consumption vary monotonously with depth, their ratio and difference reach an extreme value at some intermediate depth.

Let current velocity V decrease linearly with depth

$$V = \frac{dX}{dt} = (9.5 \cdot 10^5 - 6.3 \cdot 10^2 z) \text{ m/year}, \qquad (4.1)$$

where z is depth in meters; X is horizontal distance; and t is time. In equation (4.1) the coefficients are selected in such a manner as to correspond as closely as possible to the observed values. Thus, the average current velocity in the layer from 0-1000 m, according to (4.1), is

$$\overline{V} = \frac{1}{1\,000} \int_0^{1000} (9.5 \cdot 10^5 - 6.3 \cdot 10^2 z)\, dz = 1.9 \text{ cm/sec}, \quad (4.2)$$

which differs little from the observed average current velocity in this layer (Seiwell, 1934).

Let us now assume that oxygen consumption varies vertically according to the following law:

$$\frac{dY}{dt} = -(0.6 - 4 \cdot 10^{-7} z^2) \text{ cm}/l \cdot \text{year}. \quad (4.3)$$

The coefficients here are also close to their observed value. According to formula (4.3), the average consumption of oxygen in the layer from 0-1000 m is 0.467 cm^3/l·year, while its observed value is 0.42 cm^3/l·year.

The vertical distribution of current velocity and of oxygen consumption is shown graphically in Fig. 26.

Combining expressions (4.1) and (4.3) and integrating the results, we obtain the following formula that describes the vertical distribution of dissolved oxygen in sea water:

$$Y = 8 - \frac{0.6 - 4 \cdot 10^{-7} z^2}{9.5 \cdot 10^3 - 6.3 \cdot 10^2 z} X. \quad (4.4)$$

The integration constant was determined on the assumption that the oxygen content on the initial vertical $X = 0$ is independent of depth and equal to 8 cm^3/l.

From expression (4.4) it follows that the oxygen minimum begins to develop and deepen at the 635-m level with distance from the initial vertical. Thus, in order for the oxygen minimum to occur at a given depth it is not necessary that horizontal motion be minimal in the vicinity of this depth.

The foregoing considerations show that the oxygen minimum at intermediate depths in sea water cannot be

17. IDENTIFICATION OF THE LAYER OF NO MOTION 123

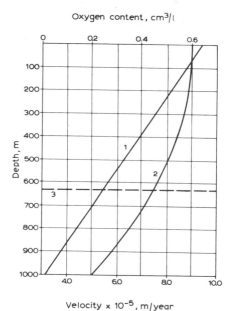

Fig. 26. Curves of the vertical distribution of current velocity and consumption of dissolved oxygen (according to Seiwell): 1—formula (4.1); 2—formula (4.3); 3—depth of the oxygen minimum.

used as the "zero" surface in the computation of sea currents by the dynamic method, because it is located at a depth where current velocity may differ from zero. The unrealistic result of dynamic computations relative to a reference surface located in the oxygen minimum is proof of this. The origin of the oxygen minimum cannot be explained from the dynamic point of view alone, because, in addition to the current, the vertical inhomogeneity of biological activity in sea water has a substantial effect on the formation and maintenance of the intermediate oxygen minimum. The pattern of vertical oxygen distribution in a sea, the ratio between its supply and consumption, and the reasons for the extremes in oxygen content can be clarified only by special biological and chemical investigations. There is no doubt, however, that the oxygen minimum does not represent the boundary between different systems of circulation. It can be located in a current and be transported by it.

18. DETERMINATION OF THE LAYER OF NO MOTION IN THE SEA FROM THE DISTORTION OF THE THICKNESS OF ISOPYCNAL LAYERS (PARR'S METHOD)

In his method of determining the layer of no motion in the sea, A. Parr (1938) tried to relate vertical current velocity distribution to the disturbance of the density field. The method is based on the assumption of isopycnal motion. It is evident that in the presence of a current the isopycnal surfaces in a cross section perpendicular to the direction of the current must slope from left to right in the Northern Hemisphere if we look in the direction of motion. Moreover, the thickness of the layer bounded by two isopycnal surfaces that we shall call pycnomere, according to Parr, cannot remain constant in the region of the current. The thickness of the pycnomere must vary in the direction perpendicular to the direction of the current. We can assume on this basis that distortion of pycnomeres in the layer with a minimum or a complete lack of water motion must be minimum or the isopycnal layers must be completely undistorted.

In Parr's opinion, the existence of an undistorted pycnomere is a sufficient condition for the absence of motion in this layer in a direction perpendicular to the cross section under consideration. Moreover, if we can prove that there is also no current along the cross section, then the depth of the undistorted pycnomere will be a reliable reference surface for computation of absolute current velocities by the dynamic method ("zero" surface).

Parr suggests that the depth of the layer of no motion in the sea (undistorted pycnomere) be determined in the following manner. The ratio of the depth increment to the corresponding increment in density $\Delta z/\Delta \sigma_t$ is computed from the vertical distribution of sea water density observed at a number of hydrological stations. In other

18. DETERMINATION OF THE LAYER OF NO MOTION 125

words, values are computed that are inverse to the vertical gradient of sea water density. Then the ratios found for all the stations that constitute the hydrological cross section under consideration are plotted on a single graph as a function of average density (Fig. 27). Since these ratios represent the thickness of layers with a single density change, the intersection or contiguity of the curves in the graph that relate to neighboring stations indicates that the thickness of the corresponding pycnomeres is the same at these two stations, i.e., the pycnomere is not significantly distorted in the interval between the two stations.

Figure 27 shows that these curves intersect at many points. The $\Delta z/\Delta \sigma_t$ curves of neighboring stations have two and sometimes more common points. According to this method, the layer of no motion is located in the region where the points of intersection of curves for neighboring stations are crowded. This corresponds to the $\sigma_t = 27.27$ isopycnal in the example under consideration. In Parr's opinion, secondary points of intersection may have a physical meaning that is not yet clear.

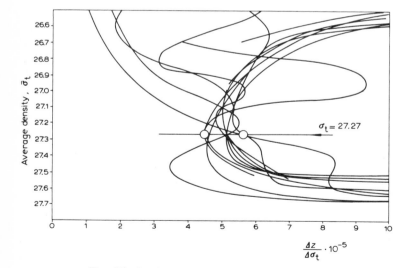

Fig. 27. Graphic illustration of Parr's method.

Parr's method is based on the principle of isopycnal water motion, and therefore the "zero" surface is considered to coincide with the isopycnal surface. In our case it is located in the region of the $\sigma_t = 27.27$ isopycnal.

Let us note that the $\Delta z/\Delta \sigma_t$ ratio is the reciprocal of the simplified expression for estimating the stability of water layers. Consequently, the layer of no motion coincides with the layer of constant stability, according to Parr. At first glance this serves as a confirmation of the reliability of the method, because increased stability interferes with the vertical transfer and intensifies the horizontal exchange of motion. At the same time, however, this fact gives rise to the most serious objections. We shall discuss this below.

Now a few words on accuracy. The accuracy of $\Delta z/\Delta \sigma_t$ computations depends on the accuracy with which density is determined. This ratio is computed with sufficient accuracy in the surface layer, where vertical density gradients are strong. The density increment between two neighboring levels of measurement is very small in deep layers. It is comparable to the accuracy of σ_t determinations. Without considering our objections at this time, we must say that Parr's method can be used to determine the boundary between current systems of different directions ("zero" surface) only if it is located at a shallow depth where the vertical density gradient is strong. The "zero" surface cannot be determined reliably by this method if it is located in a layer of weakly stratified water.

Parr suggests the following operations for the final determination of the depth of the layer of no motion. First the water masses are analyzed with respect to T-S curves and their boundaries are plotted on the cross section. Then relative salinity in the cross section is computed and the relative 50% isohalines are plotted. Thereafter the direction of the current between pairs of hydrological stations is determined from the sign of the slope of isopycnal surfaces and from the $\Delta z/\Delta \sigma_t$ ratios at neighboring stations. Finally a smoothed curve, i.e., the "zero" surface determined by Parr's method, is plotted on the cross section with consideration of all the plotted boundaries and the location of the least distorted pycnomere.

18. DETERMINATION OF THE LAYER OF NO MOTION

It should be noted that these additional operations (analysis of water masses with respect to T-S curves, determination of the location of the relative 50% isohaline and that of the direction of the current from the slope of the isopycnal layers) do not introduce significant corrections in the determination of the layer of no motion by Parr's method. According to this method, the "zero" surface coincides with the least distorted pycnomere.

Parr used his method to determine the "zero" surface in the dynamic analysis of measurements by the *Atlantis* along the profile from the Chesapeake Bay to Bermuda. The "zero" surface determined by this method is located at a depth of some 300 m in the left part of the profile, drops sharply at the line of the Gulf Stream, and, according to Parr, reaches a depth of about 2000 m in the right (oceanic) part of the profile. The author of the method believes that the location of the layer of no motion is most reliably determined in the left part of the profile, while in the right part the relation between the isopycnals and the current is less definite. The "zero" surface to the right of the periphery of the Gulf Stream is conveniently identified with the 2000-decibar isobaric surface. To confirm the results obtained, Parr compares the determined "zero" surface with the intermediate oxygen minimum and obtains almost complete identity. The oxygen minimum and the least distorted pycnomere agree over a considerable part of the profile. There is disagreement only in the right part of the profile, where the accuracy of $\Delta z / \Delta \sigma_t$ computations does not permit reliable determination of the location of the least distorted isopycnal layer.

Let us now examine this method critically. As we mentioned earlier, the vertical variation of current velocity depends not only on the slope of isopycnal surfaces, but also on the vertical water density gradient. The latter probably has the greater influence on the attenuation of current velocity with depth. Observations show that in regions where vertical density gradients are strong the slope of isopycnal surfaces and the distortion of the thickness of isopycnal layers are insignificant, whereas current velocity changes strongly along the vertical. The slope of isopycnal surfaces and the distortion of pycnomeres are considerable in layers that have a weak vertical stratification, but current velocity, usually, varies little along the vertical.

Parr's method does not consider the vertical density gradient. Thus if there is a layer with a strong vertical density gradient at some intermediate depth in the sea, the slopes of isopycnal surfaces and the distortion of pycnomeres in this layer will be minimal as compared to the overlying and underlying layers, even if current velocities are high. The intermediate maximum of the vertical water density gradient is equivalent to the intermediate maximum of the vertical stability of water layers. As a result, the least distorted pycnomere coincides with the oxygen minimum, because the rate at which oxygen-consuming organic matter sinks decreases in this layer.

Consequently, the various arguments advanced by Parr in favor of the reliability of his method represent various aspects of the manifestation of a single phenomenon, i.e., the intermediate maximum of the vertical sea water density gradient.

Let us return to Parr's example of the determination of the layer of no motion along the profile of the *Atlantis* from Bermuda to the Chesapeake Bay. The distribution of density along this profile was shown in Fig. 9. As we can see, the intermediate maximum of the vertical density gradient is clearly evident in the right oceanic part of the profile. It remains in the region of the isopycnal surfaces $\sigma_t = 27.30 + 27.40$ along the entire profile. The least distorted pycnomere determined by Parr approximates this layer, as do the intermediate maximum of vertical stability and the minimum content of dissolved oxygen. This suggests the conclusion that minimum distortion of an isopycnal layer in the sea is not a sufficient condition for the existence of a boundary between different circulation systems at this depth nor for the absence of water motion. This is because, even in the presence of a current in this layer, the strong vertical density gradient decreases the change in the thickness of the pycnomere as compared with the overlying and underlying isopycnal layers.

Parr's remark that the maximum vertical stability of water layers interferes with the vertical exchange of the amount of motion and promotes horizontal exchange is correct but cannot serve as proof of the reliability of his method, because the purpose is to determine the lower boundary of the steady gradient current. As we know, a steady gradient current is sustained by the horizontal pressure gradient, while the effect of frictional forces on this component of current velocity is insignificantly small in the stationary state.

In concluding our discussion of the method of determining the layer of no motion in the sea from the distortion of the thickness of isopycnal layers, we must note once more that the assumption of least movement in the region of least distorted pycnomeres on which this method is based is not strict. An undistorted isopycnal layer is a necessary but insufficient condition for the existence of a layer of no motion in the sea. If there is a layer with a strong vertical pressure gradient at intermediate depth, then the isopycnal layers will be least distorted as compared with the overlying and underlying layers in the presence of a strong gradient current in this layer. In seas that do not have a pronounced intermediate vertical density gradient maximum, the selection of an undistorted or slightly distorted isopycnal layer by the method under discussion becomes a very uncertain task. This will be demonstrated in more detail in the following chapter when we analyze the vertical structure of the gradient current in the region of the Kurile-Kamchatka basin in the Pacific.

19. DETERMINATION OF THE DEPTH OF THE LAYER OF NO MOTION FROM SALINITY DISTRIBUTION (HIDAKA'S METHOD)

The current velocity field in the sea is in constant interaction with the fields of any physical-chemical property of sea water. Water motion is accompanied by turbulent exchange that affects the distribution of water properties. A change in properties leads in turn to a change in water density distribution and to a change in the current velocity field in the sea. In other words, there is mutual adjustment between the current velocity field and the fields of the physical-chemical properties of sea water.

This fact was used by the Japanese oceanographer, K. Hidaka (1949), to develop a method of determining the layer of no motion in the sea ("zero" surface) from the salinity distribution.

The differential equation governing the salinity distribution S in the sea is:

$$\frac{dS}{dt} = k_1 \frac{\partial^2 S}{\partial x^2} + k_2 \frac{\partial^2 S}{\partial y^2} + k_3 \frac{\partial^2 S}{\partial z^2}, \quad (4.5)$$

where t is time and k_1, k_2, and k_3 are the coefficients of turbulent diffusion along the coordinate axes X, Y and Z respectively.

Assuming motion to be stationary and relating equation (4.5) to the layer of no motion, we obtain

$$k_1 \frac{\partial^2 S}{\partial x^2} + k_2 \frac{\partial^2 S}{\partial y^2} + k_3 \frac{\partial^2 S}{\partial z^2} = 0, \quad (4.6)$$

instead of the expression written earlier.

We can probably assume that the coefficient of horizontal turbulent diffusion does not depend on the selection of direction, i.e., $k_1 = k_2$.

Depending on whether horizontal or vertical turbulent diffusion predominates in the sea, K. Hidaka simplifies equation (4.6) by neglecting the small terms. This results in two equations which are used to determine the depth of the layer of no motion by numerical computation:

$$\frac{\partial^2 S}{\partial x^2} + \frac{\partial^2 S}{\partial y^2} = 0;$$
$$\frac{\partial^2 S}{\partial z^2} = 0. \quad (4.7)$$

The first expression in (4.7) is used for computations in regions where horizontal salt diffusion predominates, and the second for computations in regions with intense vertical exchange.

As an example of the application of his method, Hidaka determined the depth of the layer of no motion in the

19. DETERMINATION OF THE DEPTH OF THE LAYER

Atlantic from the results of observations during the *Meteor* expedition in the region from 40°S lat. to 20°N lat.

Numerical determination of $\frac{\partial^2 S}{\partial x^2} + \frac{\partial^2 S}{\partial y^2}$ values at a number of depths did not yield the expected results: $\frac{\partial^2 S}{\partial x^2} + \frac{\partial^2 S}{\partial y^2}$ differed from zero at all levels. This led Hidaka to the conclusion that horizontal salt diffusion was small in this region. Numerical determination of the second derivative of salinity along the vertical coordinate at many hydrological stations made it possible to construct a relief chart of a surface in which $\frac{\partial^2 S}{\partial z^2} = 0$. Hidaka believes that this surface corresponds to the layer of no motion in the region under study. Its relief is rather complex and it is located in the depth interval from 900 to 1400 m. Computation showed that at some points $\frac{\partial^2 S}{\partial z^2}$ differs from zero along the entire vertical. This is attributed by the author to simultaneous intense horizontal and vertical salt diffusion.

Let us briefly discuss the principles of the method in order to determine whether the "zero" surface determined by Hidaka coincides with the layer of no motion in the sea, or whether it has some other meaning.

Hidaka believes that the coefficients of turbulent diffusion remain finite in the presence of motion. This is very doubtful. Our present knowledge concerning turbulent processes in the sea is very limited, but we know that the intensity of turbulent exchange depends on the current velocity field. It is meaningless to speak of turbulent diffusion in the absence of motion.

The coefficient of turbulent diffusion (exchange) in the sea is directly related to the pulsation of current velocity. According to L. Prandtl(1951), the coefficient of turbulent friction depends on the free path of particles and on the current velocity gradient. Velocity pulsations and the length of the free path of particles are zero in the absence of motion and the coefficients of turbulent exchange are equal to zero.

Consequently, there is reason to believe that the coefficients of turbulent diffusion in the layer of no motion do not remain finite and the relationships

$$k_1 \frac{\partial^2 S}{\partial x^2} + k_1 \frac{\partial^2 S}{\partial y^2} = 0 \text{ and } \frac{\partial^2 S}{\partial x^2} + \frac{\partial^2 S}{\partial y^2} = 0$$

do not follow one from the other.

What, then, is being determined from the distribution of sea water salinity by Hidaka's method? The second derivative of the function is equal to zero when the values of the function change linearly along the selected direction or when the curve of the function changes sign (at the inflection point), and also when the values of the function remain constant along this direction. Thus, in considering vertical salt diffusion and neglecting horizontal diffusion, by Hidaka's method we determine the depth of the layer in which salinity is constant or varies linearly, or the depth of the upper and lower boundaries of the intermediate salinity extreme. It does not follow directly from this method that there is no motion in the layer corresponding to such structural characteristics of the salinity field. A positive answer to the question requires some additional arguments.

It is easy to show why Hidaka was not able to discover a surface in the ocean where the sum of the second derivatives of salinity along the horizontal coordinate axes is zero.

Let us direct the Y-axis along a tangent to the isohaline, and the X-axis along the normal to the isohaline. Then

$$\frac{\partial^2 S}{\partial x^2} + \frac{\partial^2 S}{\partial y^2} = \frac{\partial^2 S}{\partial n^2}, \qquad (4.8)$$

where n is the direction of the normal to the $S(x, y) =$ const curve.

From (4.8) it follows that the sum of the second derivatives of salinity is zero when salinity varies linearly or remains constant in some area of the horizontal plane, and also at the boundary of the extreme value of S. Apparently such conditions occur rarely over large areas of the ocean, and this led Hidaka to the wrong conclusion about the insignificance of horizontal diffusion as compared to vertical diffusion.

Hidaka's method of determining the layer of no motion in the sea ("zero" surface) cannot be considered the least reliable for these reasons. This method can be used to determine the location of a surface that corresponds to very definite structural features of the salinity field. But these characteristics of the salinity distribution are probably not uniquely related to the current velocity field. Moreover, the penetration depth of the gradient current in temperate latitudes varies from season to season; therefore, the field of any physical-chemical water property, including salinity, by retaining the features of preceding periods cannot describe the current existing at the given time.

20. COMPUTATION OF THE VERTICAL DISTRIBUTION OF CURRENT VELOCITY BY THE DYNAMIC METHOD USING CONTINUITY EQUATIONS (HIDAKA'S METHOD)

The dynamic method makes it possible to compute more or less reliably the difference in current velocities at two depths from the distribution of sea water density. To reduce the computed current velocity values to their absolute magnitudes, K. Hidaka (1940a, 1940b, 1950) suggests the use of continuity equations. Let us describe this method.

Let us examine a tetrahedral prism that extends from the sea surface to the bottom. This prism is represented by a quadrangle $abcd$ (Fig. 28) at the sea surface at whose apexes are four hydrological stations with a vertical distribution of water temperature and salinity that is known from observation. Exchange of the water volume and mass of salt takes place through the lateral faces of the prism. There is no liquid flow through the sea surface and through the bottom. Taking into consideration the six faces shown in Fig. 28 that form three individual trihedral prisms, we can write six independent equations on the basis of the continuity of the water volume and the mass of salt. These equations have the form:

$$\sum_{1}^{3} L_i \int_{0}^{h_i} v_i(z)\,dz = 0, \quad \sum_{1}^{3} L_i \int_{0}^{h_i} v_i(z) S_i(z)\,dz = 0, \quad (4.9)$$

where L_i is the distance between two hydrological stations; h_i is the average depth between the two stations; $v_i(z)$ is average current velocity; and $S_i(z)$ is the average salinity at depth z in the inverval between the two stations.

Current velocity between stations relative to the current velocity c_i at some level (for instance, the sea surface) can be computed by the dynamic method from the distribution of water density at the four stations a, b, c, and d.

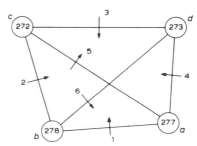

Fig. 28. Distribution of hydrological stations.
The arrows show directions that are considered to be positive.

According to this method, the absolute velocity $v(z)$ at depth z is

$$v_i(z) = u_i(z) + c_i. \qquad (4.10)$$

Here c_i is current velocity at the sea surface and $u_i(z)$ is the current velocity computed by the dynamic method relative to the velocity of the surface current. The directions marked by arrows in Fig. 28 are positive for c_i.

Substituting (4.10) in (4.9), we obtain a closed system of equations with six unknown c_i ($i = 1, 2, ..., 6$):

$$\begin{aligned}
A_1 c_1 + A_2 c_2 - A_5 c_5 + M_1 + M_2 - M_5 &= 0; \\
A_3 c_3 + A_4 c_4 + A_5 c_5 + M_3 + M_4 + M_5 &= 0; \\
A_1 c_1 + A_4 c_4 + A_6 c_6 + M_1 + M_4 + M_6 &= 0; \\
H_1 c_1 + H_2 c_2 - H_5 c_5 + N_1 + N_2 - N_5 &= 0; \\
H_3 c_3 + H_4 c_4 + H_5 c_5 + N_3 + N_4 + N_5 &= 0; \\
H_1 c_1 + H_4 c_4 + H_6 c_6 + N_1 + N_4 + N_6 &= 0.
\end{aligned} \qquad (4.11)$$

20. COMPUTATION OF THE VERTICAL DISTRIBUTION 135

These equations contain the following designations:

$$A_i = L_i h_i; \quad H_i = L_i \int_0^{h_i} S_i(z) \, dz;$$

$$M_i = L_i \int_0^{h_i} u_i(z) \, dz; \quad N_i = L_i \int_0^{h_i} u_i(z) S_i(z) \, dz; \quad (i = 1, 2, \ldots, 6).$$

(4.12)

Having solved the system of equations (4.11), we have determined the velocity of the surface gradient current in the intervals between hydrological stations. Substituting the values for c_i in expression (4.10), we find the vertical distribution of the components of current velocity that are normal to the faces of the prism.

Hidaka has computed the vertical structure of the gradient current by this method from the material of four hydrological stations of the *Meteor* located between 17° and 19°N lat. and 19° and 23°W long. (stations 272, 273, 277, and 278). The location of these stations is shown in Fig. 28. The results of his computation show the layer of no motion in this region as located between 2000 and 2500 m, and current velocities at the bottom as of the order of 10-15 cm/sec.

The simplified condition of continuity used by Hidaka in his method gives rise to a fundamental objection (Defant, 1941b). As we know, the condition of continuity requires constant mass and not constant volume. In its general form the continuity equation is

$$\frac{\partial \rho}{\partial t} + \frac{\partial \rho u}{\partial x} + \frac{\partial \rho v}{\partial y} + \frac{\partial \rho w}{\partial z} = 0.$$

(4.13)

Under stationary conditions, instead of using equation (4.9) we should write

$$\sum_1^3 L_i \int_0^{h_i} \rho_i(z) v_i(z) \, dz = 0;$$

$$\sum_1^3 L_i \int_0^{h_i} \rho_i(z) v_i(z) S_i(z) \, dz = 0.$$

(4.14)

136   IV. METHODS FOR COMPUTING THE "ZERO" SURFACE

Using the simplified continuity equation, Hidaka neglects the change in density in the volume under consideration, although this change is of the same order of magnitude as the change in salinity. Variations of both are parallel, as a rule.

It should be noted that in computations based on the application of the continuity equation to finite volume, we must deal with small differences in large values. Therefore, errors arising in the result both from the inaccuracy of initial values and from the seeming neglect of secondary processes must be estimated very strictly. In this case the neglect of salt diffusion in the prism affects the reliability of the result considerably. The same can be said of the change in density in a fixed finite volume.

As A. Defant has rightly noted, the system of equations (4.11) cannot be solved with the existing accuracy of measurements at sea. The first three equations of this system differ from the last three only by the multiplier $S_i(z)$ that enters under the sign of the integral. But salinity in the prism varies within very small limits, so that substitution of the average value of salinity in the prism yields an error of the order of 1% in the computation of the coefficient. The accuracy of the determination of distances between hydrological stations L_i and of the depth h_i at these stations is approximately 1-2%. Consequently, the differences between the coefficients of the first three equations in system (4.11) and the corresponding coefficients of the last three equations of this system do not exceed the limits of accuracy of the computation of these coefficients. Thus equations (4.11) are practically inconsistent and cannot be solved.

Hidaka obtained results in the example illustrating his method only because an unfortunate error was introduced into his computations. Instead of salinity he used data on the chlorinity of sea water, substituting, in order to facilitate computation, values for chlorinity Cl by (Cl - 18) differences, which cannot be done. Let us demonstrate this by an example. Let us rewrite the fourth equation in the system (4.11) by substituting the (Cl - 18) value for Cl.

20. COMPUTATION OF THE VERTICAL DISTRIBUTION

$$L_1 \int_0^{h_1} (\text{Cl}-18)\,dz + \underline{18L_1h_1} + L_2 \int_0^{h_2} (\text{Cl}-18)\,dz + \underline{18L_2h_2} -$$

$$L_5 \int_0^{h_5} (\text{Cl}-18)\,dz - \underline{18L_5h_5} + L_1 \int_0^{h_1} (\text{Cl}-18)\,udz +$$

$$\underline{L_1 \int_0^{h_1} 18udz - L_5 \int_0^{h_5} (\text{Cl}-18)\,udz - L_5 \int_0^{h_5} 18udz} +$$

$$L_2 \int_0^{h_2} (\text{Cl}-18)\,udz + \underline{L_2 \int_0^{h_2} 18udz} = 0.$$

(4.15)

From (4.15) it follows that reduction of sea water chlorinity by 18 units is equivalent to neglecting the underlined terms in (4.15). Such "simplification" led Hidaka to completely wrong results. When using continuity equations the initial data can be changed by several factors, but it is inadmissible to compute from them or to add a constant quantity to them.

Having the original data (Boehnecke, 1932), we repeated Hidaka's computation, taking his error into consideration. As a result we obtained the following system of equations for the determination of current velocity at the sea surface:

$$835.6747\,c_1 + 738.6832\,c_2 - 1216.0894\,c_5 - \quad 670.5633 = 0;$$
$$924.6103\,c_3 + 546.2138\,c_4 + 1216.0894\,c_5 - \quad 296.3325 = 0;$$
$$835.6747\,c_1 + 546.2138\,c_4 + 955.3405\,c_6 - \quad 154.0852 = 0;$$
$$835.3432\,c_1 + 738.5989\,c_2 - 1215.7258\,c_5 - \quad 656.9919 = 0;$$
$$924.9378\,c_3 + 546.2045\,c_4 + 1215.7258\,c_5 + 71.5608 = 0;$$
$$835.3432\,c_1 + 546.2045\,c_4 + 954.6736\,c_6 + 27020.7270 = 0.$$

(4.16)

Here the last three equations are divided term by term by the average sea water chlorinity in the tetrahedral prism under consideration ($\overline{\text{Cl}} = 19.42\%_0$).

If we consider the accuracy of computation of the coefficients, equations (4.16) must be considered as incompatible, since

$$A_1 - H_1 = 0.3315; \qquad A_4 - H_4 = 0.0093;$$
$$A_2 - H_2 = 0.0843; \qquad A_5 - H_5 = 0.3636;$$
$$A_3 - H_3 = -0.3275; \qquad A_6 - H_6 = 0.6669$$

and the determinant of the system is very close to zero. We obtained the same results by computing the vertical distribution of the gradient component of current velocity by Hidaka's method in the northwestern part of the Pacific Ocean (Chapter 6).

Thus, the method proposed by Hidaka for computing the vertical structure of the gradient current by the dynamic method using continuity equations cannot be used for practical computations, since it is not strictly theoretical and also because it leads to a system of equations that cannot be solved with the existing accuracy of measurements of physical-chemical properties of sea water.

Some authors suggest use of the known fact that in the steady state water discharge through any cross section of the ocean or sea is zero to estimate the depth of the layer of no motion. This cross section must evidently extend from one shore to the other.

Thus, Sverdrup and others (1942) recommend that the "zero" surface in the sea be determined by comparing water discharges computed by the dynamic method from a horizontal reference surface that is successively placed at different depths. The reference surface will be in the layer of no motion when water discharge through the section above the reference surface is equal to the water discharge below this surface. G. Riley (1951) uses the same method; however, in contrast to Sverdrup, he does not take the reference surface to be horizontal, but as coinciding with the surface with equal temperature or salinity.

When such use is made of the condition of continuity in the determination of the layer of no motion, it is very important that hydrological stations be close to each other along the profile and accurately represent density distribution in the selected cross section. The fact is that water moves in the ocean primarily in a horizontal plane. Currents that compensate each other are not located one above the other, as a rule, but in the vicinity of each other in the same horizontal plane. Moreover, gradient currents penetrate to considerable depths and the velocity of deep currents is usually small, but in computing the balance the small velocity values of deep currents may yield a large discharge, because the layer embraced by a deep current is sufficiently thick. Therefore, when computing water transport in one or the other direction through the plane of the cross section, the accuracy of the water density determination may prove to be insufficient, especially in deep layers. Aside from this, when the discontinuity condition is used to estimate the depth of the "zero" surface along the profile, it is highly desirable to consider the pure drift component of the velocity of the wind-driven current that comprises approximately one-tenth of the total full current along the vertical. This complicates computation considerably, however, and introduces additional mistakes because of the uncertainty of the wind field that corresponds to the stationary case.

As far as Riley's assumption of the form of the lower boundary of the current is concerned, there is no foundation, in our opinion, for the a priori premise that the "zero" surface must coincide with some real surface in the sea (isothermal or isohaline). With our present knowledge about the structure of deep currents it is convenient to use a horizontal reference surface, because in this case current velocities at various depths are computed with an approximately constant relative mistake over the entire water area. This has already been noted in the preceding chapter.

21. DETERMINATION OF THE DEPTH OF THE "ZERO" SURFACE FROM THE ANALYSIS OF DIFFERENCES IN THE DYNAMIC DEPTHS OF ISOBARIC SURFACES (DEFANT'S METHOD)

In examining the differences in the dynamic depths of isobaric surfaces at a great number of pairs of neighboring stations, A. Defant (1941a) discovered a layer of relatively great thickness in which these differences vary considerably along the vertical. The thickness of this layer in the Atlantic Ocean ranges from 300 to 800 m and its depth varies rather uniformly in the horizontal direction, while the change in differences in the dynamic depths of isobaric surfaces only amounts to several dynamic millimeters.

The constancy of differences in dynamic depths indicates, as we know, that the gradient component of current velocity is constant along the vertical in this layer. Defant assumes that this water layer is almost motionless and considers it to be the layer directly adjoining the "zero" surface. He illustrates his arguments with an example whose graphical representation is given in Fig. 29. As we can see from the graph, the curve for differences in dynamic depths between the 1000- and 1600-decibar levels is almost vertical. Defant suggests that this layer be considered the layer of no motion.

Of all the existing methods for determining the depth of the layer of no motion in the sea, Defant's method is the most justified. Indeed, the curve for the vertical distribution of differences in the dynamic depths of isobaric surfaces is identical with the curve for the velocity component of the gradient current that is normal to the profile, or, more accurately, the curve of current velocity relative to the velocity of the gradient current at the surface. It is sufficient to multiply the difference in dynamic depths by $(2\omega L \sin \varphi)^{-1}$ to obtain the velocity of the gradient current with an accuracy to the constant value. This constant value is equal to current velocity at the sea surface. Consequently, Defant's method of determining the "zero" surface is based on analysis of the vertical variation of the velocity of the current being computed by the dynamic method. Defant believes that a weak vertical current velocity gradient testifies to the insignificant value of velocity itself.

In the author's opinion, the method of identifying the layer of no motion with the layer where the differences in dynamic depths are similar is most justified and plausible. Any other assumption concerning the location of the layer of no motion (Fig. 29) is arbitrary and makes the presence of almost constant current velocity in the 1000- to 1600-decibar layer obscure. If we assume that the layer with similar differences in dynamic depths is the boundary between currents of varying directions, then weak vertical current velocity gradients must evidently occur in this layer.

Such arguments, however, cannot be considered as proof of the absence of a current or of minimum current velocity in the layer where differences in dynamic depths are constant. Substantiation or refutation of Defant's hypothesis requires additional valid arguments. For instance, it is

21. DETERMINATION OF THE DEPTH OF THE "ZERO" 141

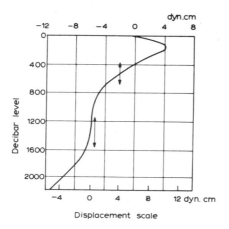

Fig. 29. Graphical illustration of Defant's method.

doubtful whether we can identify the layer with similar differences with the layer of no motion when the current computed by the dynamic method has the same direction above and below the "zero" surface. Moreover, it is hardly probable that in deep ocean layers in the immediate vicinity of the bottom there are currents that are opposite in direction to surface currents and velocities that frequently exceed the velocity of gradient currents at the surface. These are the results that are obtained when deep currents are computed by the dynamic method relative to a "zero" surface, as defined by Defant's method (Wüst, 1955).

The strict constancy of differences in the dynamic depths of isobaric surfaces at some depth interval means that the increments in dynamic depths between two levels are equal at two neighboring stations, i.e.,

$$\Delta D_A = \Delta D_B \tag{4.17}$$

or

$$\left[\int_{p_n}^{p_{n+1}} \alpha dp \right]_A = \left[\int_{p_n}^{p_{n+1}} \alpha dp \right]_B. \tag{4.18}$$

In these expressions ΔD_A and ΔD_B are the increments in dynamic depths between the isobaric surfaces p_n and p_{n+1} at stations A and B, and α is the specific volume of sea water.

Geometrically, equation (4.18) means that the areas formed by the curves for the vertical distribution of specific volume between surfaces p_n and p_{n+1} are equal at stations A and B. Hence, there is at least one depth in the $[p_n, p_{n+1}]$ interval where the specific volume of water is the same at stations A and B. If the distance between hydrological stations is small, there is at least one horizontal isostere in the $[p_n, p_{n+1}]$ interval and the slopes of the isosteric surfaces must be opposite in sign above and below it.

The same conclusion can be obtained in another way. Recalling expression (1.9) obtained earlier, we have

$$\int_z^H \frac{\partial \rho}{\partial x} dz = \frac{\bar{\rho}}{g} \frac{\partial D}{\partial x}.$$

According to Defant's method, the difference in the dynamic depths of isobaric surfaces is constant along the vertical in the layer where current velocity is zero. Let us differentiate the above equation with respect to z and equate the result to zero

$$\frac{\partial \rho}{\partial x} = \frac{\bar{\rho}}{g} \frac{\partial^2 D}{\partial x \partial z} = 0. \qquad (4.19)$$

From formula (4.19) it follows that $\rho(x) = $ const in this layer. In order to determine the "zero" surface by Defant's method, it is sufficient to analyze the distribution of density (or specific volume) along the vertical cross section and to mark the depths at which the horizontal density gradient changes sign. As a rule, there is a large number of areas in the cross section where $\frac{\partial \rho}{\partial x} = 0$ and it is very difficult to draw a smooth curve. This is partially due to measurement inaccuracies, because frequently the horizontal variation of density does not exceed the limit of accuracy of density computation. Indeed, the horizontal variation of density in the deep layers between two stations often does not exceed ±0.02 to ±0.04 σ_t units, whereas the accuracy of measurement of the difference in density is equal to the same values (Chapter 2).

A. Defant applied his method of determining the layer of no motion in the sea to the analysis of a large number of observations in the Atlantic. Layers where differences in dynamic depths varied only slightly along the vertical were plotted at all stations. Then a smoothed curve crossing the centers of layers with similar differences plotted for each pair of stations was drawn along each profile. Defant suggests that this smoothed curve be considered as the "zero" surface for the computation of absolute current velocity by the dynamic method.

Figure 30 shows the vertical distribution of differences in the dynamic depths of isobaric surfaces along the meridional profile of the Atlantic. The arrows designate layers with similar differences in dynamic depths. As we can see from the graph, the selection of the depth of the layer of no motion is somewhat uncertain. However, if we consider that the curves were selected from extensive material to illustrate the method as convincingly as possible, it becomes clear to what extent the selection of the layer with similar differences in dynamic depths depends on the individual approach and subjectivity of the investigator. In some cases there are several layers with a similar difference in dynamic depths, and to give preference to one of these seems arbitrary.

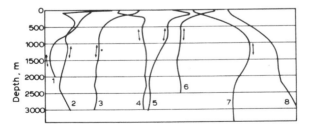

Fig. 30. Curves for the differences in the dynamic depths of isobaric surfaces between pairs of stations along the meridional profile in the Atlantic Ocean (according to Defant):
1— $\varphi = 8°$ N; 2— $\varphi = 31.7°$ N; 3— $\varphi = 11°$ N; 4— $\varphi = 0.5°$ N; 5— $\varphi = 7.7°$ S; 6— $\varphi = 18.6°$ S; 7— $\varphi = 30.3°$ S; 8— $\varphi = 48.7°$ S.

144 IV. METHODS FOR COMPUTING THE "ZERO" SURFACE

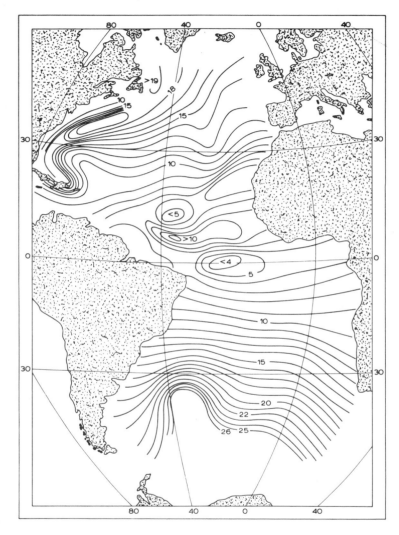

Fig. 31. Relief of the "zero" surface in the Atlantic Ocean (according to A. Defant).

Having analyzed a great number of observations made by various ships in the Atlantic, Defant determined the

21. DETERMINATION OF THE DEPTH OF THE "ZERO" 145

depth of the "zero" surface by the above method and constructed a chart of its relief (Fig. 31). As we can see from the diagram, the "zero" surface is located at depths ranging from 400 to 2600 m. Its depth is least in the vicinity of the equator and increases with increasing latitude. This corresponds rather well with theory. We mentioned earlier that the depth of penetration of a wind-driven current (thickness of the baroclinic layer) depends on the geographical latitude of the area, on salinity, and on the curl of the field of tangential wind stress. Our knowledge of the distribution of these values over the area of the Atlantic Ocean verifies in general outline the relief of the "zero" surface obtained by Defant. The "zero" surface in the South Atlantic is deeper than at corresponding latitudes of the North Atlantic because the vertical inhomogeneity of the water is less pronounced and the winds are stronger in the Southern Hemisphere.

G. Neumann (1942) determined the depth of the "zero" surface in the Black Sea using Defant's method, and then computed currents at the surface by the dynamic method. Figure 32 shows the vertical distribution of differences in the dynamic depths of isobaric surfaces at several pairs of hydrological stations, and Fig. 33 gives the relief of the "zero" surface in the Black Sea according to Neumann. Even though the scheme of surface water circulation obtained by Neumann agrees with the existing views concerning currents in the Black Sea, the reliability of "zero" surface determinations from the results of observations that rarely reached a depth of 300 m and were mostly limited to a depth of 100-200 m is doubtful. Even the few curves for dynamic depth differences in Fig. 32 show that the layer with similar differences is not always reliably selected. Some station pairs do not exhibit similar differences within the layer under consideration.

A. D. Dobrovol'skii (1949) analyzed a number of profiles in the Pacific Ocean by Defant's method and did not obtain such good results as were obtained by the author of the method in the Atlantic Ocean. The profiles exhibited two layers with relatively uniform differences in dynamic depths. A. D. Dobrovol'skii indicates that these layers were determined in the conventional manner. At some pairs of stations there were no similar differences in dynamic

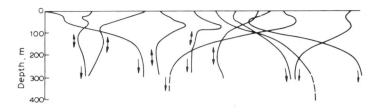

Fig. 32. Curves for the differences in the dynamic depths of isobaric surfaces between pairs of stations in the Black Sea (according to G. Neumann).

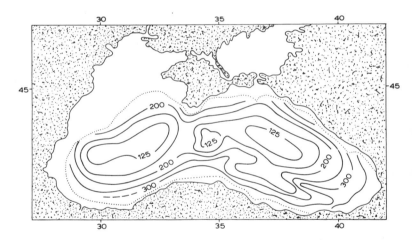

Fig. 33. Relief of the "zero" surface in the Black Sea (according to G. Neumann).

depths, whereas at other pairs of stations ΔD was relatively constant. Comparison of current velocities computed by the dynamic method with observed velocities showed that their agreement improves with increasing depth of the "zero" surface. We may thus conclude that in this example the velocity of the current in the layer with relatively uniform differences in the dynamic depths of isobaric surfaces is apparently remote from zero.

21. DETERMINATION OF THE DEPTH OF THE "ZERO"

The negative results obtained in the determination of the "zero" surface in the Pacific by Defant's method do not mean that the method is unsuitable for practical purposes. Its main advantage is that it analyzes the vertical distribution of the velocity of the current being computed by the dynamic method. If we adopt the hypothesis that current velocity is insignificant in the layer where the vertical current velocity gradient is least, then the disadvantage of the method consists only in the absence of a criterion for the uniformity of differences in the dynamic depths of isobaric surfaces. Therefore, in those cases where there is no clearly defined layer with constant dynamic depth differences, it is convenient to reduce ΔD values to current velocities relative to the velocity of the surface current, i.e., multiply ΔD by $M = (2\omega L \sin \varphi)^{-1}$. In doing this we make the results comparable when the distance between hydrological stations varies, and it becomes possible to use a single criterion for the homogeneity of current velocity. A given small vertical current velocity gradient can serve as this criterion.

When Defant's method of determining the layer of no motion in the sea is used, it should not be forgotten that current velocity is computed with a low accuracy by the dynamic method because of the accumulation of errors. If current velocity is low in the given region, Defant's method will not yield positive results, since the computed dynamic depth differences will be comparable to the probable computation error.

Let us now discuss briefly the hypothesis on which Defant's method is based. Strictly speaking, a low vertical current velocity gradient does not indicate whether there is little or no water motion in the given layer. If such layer is located in weakly stratified, almost homogeneous water, the small change in velocity along the vertical does not seem unnatural, since weakly stratified water cannot affect the horizontal pressure gradient noticeably and the gradient component of current velocity cannot change substantially along the vertical coordinate. But the situation changes radically when the layer with constant differences in the dynamic depths of isobaric surfaces is located in stratified water. The mutual adjustment of the density and current velocity fields is unthinkable from the mechanical point of view without vertical change of current velocity in a stratified water mass. Therefore, Defant's assumption is most justified in this case. The layer where current velocity does not change vertically can be considered as a layer without current. Of course, it is assumed here that the layer of constant differences is determined after consideration of the accuracy of measurements and computation errors and that in this respect there is no doubt about the location of the layer of constant differences.

There is another method of determining the depth of the layer of no motion in the sea that is similar in principle to Defant's method but differs from it technically (Mamaev, 1955). The difference consists in that the curves for the vertical distribution of differences in the specific water volume α at two neighboring stations are analyzed, instead of the differences in the dynamic depths of isobaric surfaces. Here the constant ΔD value in the layer of no motion corresponds to a zero $\Delta \alpha$. The advantage of this modernization consists in that the $\Delta \alpha$ minimum is more easily found than the layer where ΔD is constant.

CONCLUSIONS

In spite of the great importance for oceanology of determination of the "zero" layer in the sea and the efforts of a number of scientists, there is as yet no universal objective method for determining the layer of no motion or a layer of least water movement in the sea.

Of the existing methods for determining the depth of the "zero" surface, the best are that of A. Defant and H. Sverdrup's method based on the continuity of the water volume. Use of these methods for practical purposes, however, requires a careful approach in each individual case.

CHAPTER V

Practical Procedures for Computing Currents by the Dynamic Method

22. COMPUTATION OF CURRENTS IN SHALLOW AREAS BY THE DYNAMIC METHOD

Difficulties are frequently encountered in the computation of currents from water density distribution when the region under study, or an individual hydrological profile, extends into the shallow part of the sea where depths are less than the depth of the selected reference surface. In such cases current velocity can be computed by the ordinary dynamic method only in the deep part of a profile or basin. To compute currents in the shallow part we must either select another reference surface and obtain results that are not comparable over the entire area, or use some artificial means to eliminate this difficulty.

We shall discuss such procedures below and evaluate each. Let us note that here we shall assume for the sake of simplicity and the sequence of arguments that the "zero" or reference surface is horizontal. In a more general case, i.e., when the form of the "zero" surface is complex, the methods of dynamic computation do not change significantly, but merely become somewhat more complicated. These methods must be supplemented by those described in the following section, where methods of computing current velocities relative to a "zero" surface of an arbitrary complex form are discussed.

FIRST METHOD. Let us assume that a hydrological profile, whose density distribution is known from observation,

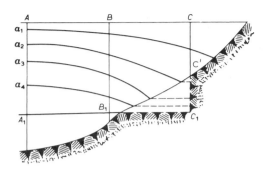

Fig. 34. Hydrological profile, supplemented by an imaginary motionless water mass.

extends from the deep part of the sea into the shallow part where depths are less than the depth of the reference surface. This profile, which is formed by three hydrological stations A, B, and C, is represented in Fig. 34. The sloping lines on the profile show the distribution of the specific volume of water a. The reference surface is horizontal and is located at the depth A_1B_1. Current velocity at the reference surface is assumed to be insignificantly small or strictly equal to zero. Current velocity can obviously be computed relative to this reference surface only in the interval between hydrological stations A and B. The reference surface between stations B and C is below the bottom line and computation relative to it is impossible.

Let us replace the block of solid earth in the triangle $B_1C'C_1$ by an imaginary water mass. If we assume that current velocity and the horizontal pressure gradient are zero along the bottom line, then the isobaric and isosteric surfaces in this imaginary water mass must be horizontal. On this basis we project horizontally the points of intersection of the isosteric lines with the bottom into the vertical of station C, as shown in Fig. 34 by dashed lines. Now we have the vertical distribution of specific volume at the three stations A, B, and C from the sea surface to the depth of the reference surface $A_1B_1C_1$ for computation of current velocities by the dynamic method.

22. COMPUTATION OF CURRENTS IN SHALLOW AREAS

The above method of computing current velocities by the dynamic method in shallow areas is usually associated with Helland-Hansen (1934), who used it to compute currents in one of the Norwegian fjords, but the first such method was described and used by Mohn (1885).

SECOND METHOD. This procedure for computing currents by the dynamic method along shallow sections of a profile is also based on the assumption that current velocity and the horizontal pressure gradient along the bottom line are zero. In this method the hydrological profile is also supplemented by an imaginary motionless water mass and it is assumed that the bottom line between two stations is rectilinear, while the isosteres in the imaginary water mass are equidistant near the bottom. This method is used in the following manner in practical computations.

Let us take two hydrological stations A and B of different depths (Fig. 35) and compute the vertical distribution of average current velocity relative to a horizontal reference surface passing through point B_0. To do this we must first compute the differences in the dynamic heights of isobaric surfaces from the reference surface $A_0 B_1$ and then add to them a correction Δ.

$$\Delta = \frac{1}{2} h (\alpha_{B_1} - \alpha_{A_0}). \tag{5.1}$$

Here h is the difference in the depths of neighboring stations expressed in pressure units, and α_{B_1} and α_{A_0} are the specific volumes of water at points B_1 and A_0, respectively.

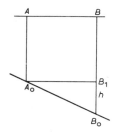

Fig. 35. Hydrological profile formed by two stations of varying depths.

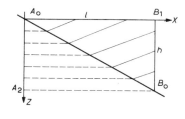

Fig. 36. Profile, supplement by an imaginary water mass in which the isopycnal layers are horizontal and equidistant.

Let us examine the derivation of the expression for the correction term (5.1). According to assumption, the isosteres must be parallel and equidistant at the bottom along the profile, and horizontal and equidistant in the imaginary water mass. Let us examine Fig. 36, which shows the section of the hydrological profile adjoining the bottom. A_0B_0 is the bottom line, l is the distance between the two hydrological stations, and h is the difference in depth at these stations. The X-axis is directed from left to right, and the Z-axis vertically downward.

If we designate the specific volume of water at point A_0 with α_0, then, according to the assumptions on which this method is based, the distribution of specific volume within triangle $A_0B_0B_1$ is described by equation:

$$\alpha = \alpha_0 + x\frac{\partial \alpha}{\partial x} + z\frac{\partial \alpha}{\partial z}, \qquad (5.2)$$

where $x = zl/h$ and $\frac{\partial \alpha}{\partial x}$ and $\frac{\partial \alpha}{\partial z}$ are constant.

The distribution of specific volume along the vertical A_0A_2 (Fig. 36) is:

$$\alpha_A = \alpha_0 + \frac{zl}{h}\frac{\partial \alpha}{\partial x} + z\frac{\partial \alpha}{\partial z}, \qquad (5.3)$$

and along the vertical B_1B_0:

$$\alpha_B = \alpha_0 + l\frac{\partial \alpha}{\partial x} + z\frac{\partial \alpha}{\partial z}. \qquad (5.4)$$

Subtracting (5.3) from (5.4) and integrating the result with respect to z from 0 to h, we obtain

$$h(\bar{\alpha}_B - \bar{\alpha}_A) = \frac{lh}{2}\frac{\partial \alpha}{\partial x} \qquad (5.5)$$

or, passing over to finite values,

$$\Delta = \frac{1}{2}h(\alpha_{B_1} - \alpha_{A_0}).$$

22. COMPUTATION OF CURRENTS IN SHALLOW AREAS

We obtained an expression representing, under the given assumptions, the difference in the dynamic heights of isobaric surfaces A_0B_1 read from the surface A_2B_0. Adding the value of the correction term Δ to dynamic height differences measured from the surface A_0B_1, we obtain values that describe current velocity at various horizons relative to the reference surface A_2B_0.

This method for the dynamic analysis of measurements in the shallow part of a profile was proposed by Jacobsen and Jensen (1926).

In contrast to the first, the second method has the advantage that its application does not require any graphical constructions. But the Jacobsen-Jensen method includes assumptions concerning the structure of the density field near the bottom (parallelism and equidistance of isosteres along the profile) that connot be considered as strict. The water density field rarely has such structure under actual conditions.

Both these methods of dynamic data analysis are based on the assumption that the velocity of the gradient current is zero near the bottom and, consequently, the horizontal pressure gradient is also equal to zero. This assumption is not justified, as a rule, in the application of these methods of computation. Indeed, such methods of dynamic analysis are used when the computed layer of minimum horizontal water motion, the "zero" surface, is at a depth that is greater than the depths in the shallow part of the region under study. Since there is no reason to expect a strong change in the vertical structure of water masses over a small area, or changes of the order of magnitude of the slope of the free surface of the sea, the horizontal pressure gradient cannot be zero at the bottom in the shallow part of a region. This means that the gradient component of current velocity at the bottom differs from zero in shallow water. As a result of fluid adhesion, the velocity of the total current at solid boundaries becomes zero owing to friction and to the appearance in the bottom layer of an additional component of current velocity (analogous to the pure drift component of the velocity of a wind-driven current at the surface) that compensates the gradient current. But since the dynamic method can be used to compute the gradient component of the velocity of a steady current only from the density distribution and cannot be used to compute velocity components due to frictional forces, the assumption on which the first two methods of computing currents in shallow water are based, that the velocity of the gradient current at the bottom is zero, is incorrect. This should be kept in mind when these methods are used in practice.

THIRD METHOD. Graphical extrapolation of the specific volume of water in the imaginary water mass along the profile can be recommended to eliminate the rough assumption of the absence of a horizontal pressure gradient at the bottom of shallow water. In this case the solid earth along the profile from the bottom line to the "zero" surface is replaced by an imaginary moving water mass, and not by a motionless water mass. But the results of dynamic analysis using this method largely depend on the arbitrariness of the investigator who performs the extrapolation and on how large a part of the hydrological profile is occupied by the rise in the bottom. Better results can be expected when the rise in the bottom is located in the center of the region under investigation and has the form of an island or submarine peak. Subjective errors will be largest if the shallow water is located at the boundary of the basin under study.

FOURTH METHOD. This procedure for computing currents in a shallow area of a hydrological profile by the dynamic method is based, as are the first three methods, on the substitution of the solid mass of earth by an imaginary water mass confined between the bottom line and a horizontal "zero" surface. P. Groen (1948), the author of the method being described, suggests that the density field be extrapolated in such manner that the isosteres on each level in the imaginary water mass have a constant slope equal to the slope of the isostere at the point where it meets the bottom line.

Figure 37, which demonstrates this method, shows the distribution of the specific volume of water. A horizontal line is drawn from the point of intersection of each isostere with the bottom line. The cross strokes on the horizontal line indicate the slope of the isosteres at the corresponding level. The isosteres are extended by dashed lines into the imaginary water mass in such manner as not to cross the horizontal lines at the angles indicated by the cross strokes.

This method of extrapolating the density field reduces subjective errors, but why the slopes of the isosteres are constant at each horizon in the imaginary water mass remains obscure.

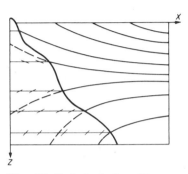

Fig. 37. Hydrological profile, supplemented by an imaginary water mass in which the slope of the isopycnal lines is constant at each horizon. The strokes indicate the angles of the isopycnal lines at their intersection with the bottom line.

All four procedures for computing currents in a shallow area of a hydrological profile by the dynamic method have one thing in common, namely, that the section of solid earth along the profile that is located above the reference surface is replaced by an imaginary water mass. In the first two procedures this mass is assumed to be motionless. In the last two methods it is assumed to be a moving baroclinic mass. The last two methods are best suited for practical use, but they are not without shortcomings.

23. COMPUTATION OF CURRENTS BY THE DYNAMIC METHOD RELATIVE TO A NONHORIZONTAL "ZERO" SURFACE

In the computation of sea currents by the dynamic method, the dynamic heights of isobaric surfaces are frequently computed relative to a horizontal reference surface that either coincides with the lower level of measurement in the region under study, or is selected on the basis of certain definite considerations. Obviously, such computations give only an approximate characteristic of the velocity field in the sea with an accuracy to the current velocity at the reference surface. The desire to obtain absolute values for the gradient component of current velocity by the dynamic method inevitably requires the determination of the layer of no motion in the sea, i.e., the "zero" surface.

Numerous measurements and computations of sea currents show that the isotachs at sea, as well as the "zero" surface and the surface where current velocity is low, are not horizontal. They are of complex form and are located in the water in conformity with the distribution of forces and factors that affect the gradient component of current velocity. This point of view does not require any proof. The extensive use of a horizontal reference surface in practical dynamic computations is dictated, in our opinion, by two circumstances: first, the lack of a reliable method of determining the "zero" surface in the sea. Existing methods have shortcomings that do not always permit the use of even the best of them. Secondly, the derivation of a dynamic method formula is based on the analysis of current velocity circulation along some closed contour whose lower boundary is assumed to be rectilinear and horizontal. Consequently, the theoretical approach itself limits the application of the dynamic method to horizontal reference surfaces.

Application of the dynamic method requires development of a reliable method of determining the layer of no motion and simplification of computations relative to a "zero" surface of arbitrary complex shape, since, generally speaking, this surface is not horizontal.

Let us examine methods for the dynamic computation of currents relative to a "zero" surface of complex shape.

FIRST METHOD. The first attempt to use a nonhorizontal reference surface in dynamic computation was made by Dietrich (1937). He proposed a method analogous to geodetic leveling. Let us take three hydrological stations A, B, and C, with a known vertical water density distribution

23. COMPUTATION OF CURRENTS

and a known "zero" surface along the profile formed by these stations (Fig. 38). It is quite clear that if the dynamic heights of isobaric surfaces are computed relative to this sloping "zero" surface and current velocities between neighboring stations are computed from the differences in these dynamic heights, the results will be incorrect.

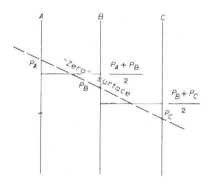

Fig. 38 Graphical illustration of Dietrich's method.

Therefore, Dietrich suggested that the sloping "zero" surface along the profile be replaced by a stepped surface consisting of individual horizontal links. The average position of the "zero" surface is found for each pair of neighboring hydrological stations as the arithmetic mean of the depths of the "zero" surface at neighboring stations. If the "zero" surface at stations A, B, and C in the example shown in Fig. 38 is located at a depth of P_A, P_B, and P_C decibars, respectively, the sloping "zero" surface is replaced by two horizontal links that are respectively located at depths

$$P_{AB} = \frac{1}{2}(P_A + P_B);$$
$$P_{BC} = \frac{1}{2}(P_B + P_C). \tag{5.6}$$

The sloping "zero" surface along a hydrological profile formed by a greater number of stations is replaced

158 V. COMPUTING CURRENTS BY THE DYNAMIC METHOD

by a step-patterned surface consisting of a larger number of horizontal links. Now each pair of hydrological stations has its own reference surface relative to which the dynamic heights of the isobars along the profile and current velocities are computed.

The operations in the dynamic computation of currents by Dietrich's method are complicated considerably when current velocities are computed from data of a hydrological survey of an ocean area rather than from data for a single profile. Construction of a dynamic isobaric surface chart requires knowledge of dynamic heights at each station reduced to comparable form, not the differences in dynamic heights between individual pairs of stations. For this purpose the dynamic heights of the isobaric surface above all the horizontal segments of averaged reference surfaces terminating at a station are computed for each hydrological station from the vertical distribution of the specific volume of water, and the differences in dynamic heights between all pairs of neighboring stations are determined. Then one of the stations is taken as the principal station.

The principal station is selected as close as possible to the center of the area, so that the vertical density distribution at this station may be determined most reliably. Then a triangulation network is constructed with hydrological stations at the vertexes. The dynamic heights of the isobaric surface at each station are found by successive addition of differences in dynamic heights between two consecutive neighboring stations to the dynamic height at the principal station. The differences in dynamic heights between stations are computed relative to an averaged horizontal "zero" surface. The triangulation network is necessary for the control of computations and the distribution of discrepancy. Let us explain it by an example.

Figure 39 shows a chain of triangles formed by hydrological stations. If we consider station 1 as the principal station and designate the difference in the dynamic heights of the isobaric surface between neighboring stations with δ_{ij}, the dynamic heights at stations 2 and 3 will be

$$D_2 = D_1 + \delta_{12};$$
$$D_3 = D_1 + \delta_{12} + \delta_{23};$$

(5.7)

23. COMPUTATION OF CURRENTS

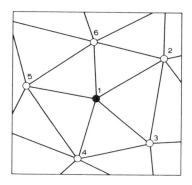

Fig. 39. Network of triangles in whose vertexes hydrological stations 1-6 are located.

where D_1 is the dynamic height of the isobaric surface at station 1.

Control is achieved by summation of the differences in the dynamic heights of the isobaric surface along a closed contour

$$\delta_{12} + \delta_{23} + \delta_{31} = 0. \tag{5.8}$$

This sum must be zero. If it differs from zero, then the discrepancy is distributed along the contour in proportion to the distance between stations. A similar operation is performed along the contour of the triangles of all stations, and the discrepancy is distributed until dynamic heights are obtained at all stations whose differences on any closed contour satisfy equation (5.8).

Defant (1941a) used this method of constructing a dynamic chart relative to a nonhorizontal "zero" surface when he studied currents in the Atlantic Ocean. He determined the location of the "zero" surface by his method from the differences in the dynamic depths of isobaric surfaces. He took station 248 of the *Meteor* ($\varphi = 3°30'$ S lat., $\lambda = 22°36'$ W long.) as the principal station and constructed a network of 1028 triangles and 629 stations. The results of the computation were good, but required cumbersome and tedious work. This is one of the great disadvantages of Dietrich's method.

SECOND METHOD. As we have mentioned earlier, the formula for computation by the dynamic method is based on the assumption that the reference surface is horizontal. Therefore, the dynamic method cannot be used in its ordinary form to compute gradient current velocities relative to a "zero" surface of complex form. M. M. Somov (1937) was the first to attempt to eliminate this discrepancy between the theoretical premises of the method and actual conditions. He derived a formula for computation by the dynamic method assuming that the reference surface is not horizontal, but slopes at some angle to the horizontal. It must be said that M. M. Somov treats his method as a means of computing currents between hydrological stations of various depths. The reference surface is considered to be the bottom line. Here we shall examine his method as a means for computing currents by the dynamic method relative to a nonhorizontal "zero" surface.

Let M and N be two hydrological stations along the profile (Fig. 40); AB, the trace of the zero surface; p_1 and p_2, the depths of the zero surface at the stations expressed in units of pressure; and L, the distance between stations. From the theory of V. Bjerknes it follows that

$$\oint_S \alpha \, dp = 2\omega \frac{\partial \Sigma}{\partial t}. \tag{5.9}$$

Here α is the specific volume of water; p is pressure; ω is the angular speed of rotation of the earth; t is time; and Σ is the projection of the $MM'N'N$ contour on the equatorial plane. The curvilinear integral in the left part of expression (5.9) is selected along the closed contour $AMNB$.

Using known transformations (Shuleikin, 1953), instead of expression (5.9) we obtain

$$v = \frac{1}{2\omega L \sin \varphi} \left(D_M - D_N + \int_{p_1}^{p_2} \alpha \, dp \right), \tag{5.10}$$

where v is the component of gradient current velocity normal to the profile at horizon p_0; D_M and D_N are the dynamic heights of the isobaric surface p_0 above the "zero" surface AB at stations M and N, respectively.

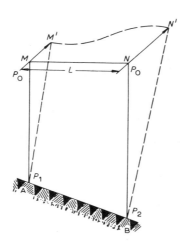

Fig. 40. Graphical illustration of the derivation of a formula for computation by the dynamic method according to M. M. Somov.

Formula (5.10) differs from the usual formula for computation by the dynamic method in that it contains an additional term in its right-hand part that we shall call, according to M. M. Somov, the "mark" of the "zero" surface. This is computed in practice by the formula:

$$\int_{p_1}^{p_2} \alpha\,dp = \frac{\alpha_1 + \alpha_2}{2}(p_2 - p_1). \tag{5.11}$$

In this expression α_1 and α_2 are the specific volumes at points A and B and at depths p_1 and p_2 decibars.

A dynamic chart of the isobaric surface is constructed in the following manner. First, one of the stations is selected as the principal station. Then the "mark" of the "zero" surface is computed at all stations relative to the principal hydrological station according to formula (5.11). The computations must also be controlled here by summation of "zero" surface marks along closed contours and by distribution of the discrepancy obtained. After all the "zero" surface marks are computed and verified, they are added to the dynamic heights of isobaric surfaces read from the "zero" surface, and the data obtained are used to construct a dynamic relief chart of isobaric surfaces.

The method proposed by M. M. Somov for constructing a dynamic chart of the isobaric surface relative to a "zero" surface of complex form has the same shortcoming as Dietrich's method, i.e., it requires tedious operations for leveling dynamic heights and reducing them to a comparable form. Therefore, let us try to find another, simpler, and, at the same time, sufficiently reliable method for dynamic computation of currents relative to a "zero" surface of an arbitrary complex form.

THIRD METHOD. Let us take a rectangular system of coordinates with the X-axis lying in the horizontal plane, and the Z-axis directed vertically downward (Fig. 41). Let us locate the origin of coordinates somewhat above the sea surface $z = \zeta(x)$. Let us assume that the "zero" surface is located at a depth $z = H(x)$. Henceforth we shall not impose any limitations on functions $\zeta(x)$ and $H(x)$, except that they be differentiable.

In the absence of frictional forces the steady motion of a fluid along the normal to the XOZ plane is defined by the relationship

$$2\omega v \sin \varphi = \frac{1}{\rho} \frac{\partial p}{\partial x}. \qquad (5.12)$$

As the second equation, let us write the hydrostatic equation

$$g = \frac{1}{\rho} \frac{\partial p}{\partial z}. \qquad (5.13)$$

In these equations ω is the angular speed of rotation of the earth; v is the component of the gradient component

23. COMPUTATION OF CURRENTS

of current velocity; φ is the geographical latitude of the area; ρ is sea water density; p is pressure; and g is the acceleration of gravity.

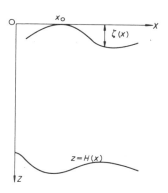

Fig. 41. Coordinate system:
$z = \zeta(x)$ —equation for the trace of the free surface of the sea; $z = H(x)$ —equation for the trace of the "zero" surface.

Having integrated expression (5.13) with respect to z and differentiated the result with respect to x, we obtain

$$\frac{\partial p}{\partial x} = g \int_{\zeta(x)}^{z} \frac{\partial \rho}{\partial x} dz - g\rho(\zeta) \frac{\partial \zeta(x)}{\partial x}. \qquad (5.14)$$

If there is no motion at depth $z = H(x)$, i.e., the "zero" surface is located at this depth, then instead of (5.14) we have

$$0 = g \int_{\zeta(x)}^{H(x)} \frac{\partial \rho}{\partial x} dz - g\rho(\zeta) \frac{\partial \zeta(x)}{\partial x} + g\rho(H) \frac{\partial H(x)}{\partial x} \qquad (5.15)$$

By combining equations (5.14) and (5.15) we find the expression for the horizontal pressure gradient at horizon z.

$$\frac{\partial p}{\partial x} = g \int_{H(x)}^{z} \frac{\partial \rho}{\partial x} dz - g\rho(H)\frac{\partial H(x)}{\partial x}. \qquad (5.16)$$

Let us now transform the integral of equation (5.16) into a form suitable for dynamic computation. Let us express water density by specific volume $\alpha = \frac{1}{\rho}$ and let us integrate with respect to the variable p. To retain uniformity in the right-hand part of (5.16) we must also transform the second term, replacing depth by pressure. Then we obtain

$$\frac{\partial p}{\partial x} = \bar{\rho}\frac{\partial D}{\partial x} - \frac{\partial h(x)}{\partial x}. \qquad (5.17)$$

In this expression $\bar{\rho}$ is average water density; D is the dynamic height of the isobaric surface p above the "zero" surface p_H:

$$D = \int_{p_H}^{p} \alpha\, dp; \qquad (5.18)$$

$h(x)$ is a function describing the depth of the zero surface in units of pressure. Henceforth we can assume that function $h(x)$ describes the depth of the layer of no motion not in units of pressure, but in units of length, because in dynamic height computations pressure in decibars and depth in meters are considered to be practically identical.

Having substituted the expression for the horizontal pressure gradient (5.17) in equation (5.12), we obtain a formula for computing the gradient component of current velocity relative to a "zero" surface of arbitrary form:

$$v = \frac{1}{2\omega \sin \varphi}\left(\frac{\bar{\rho}}{\rho}\frac{\partial D}{\partial x} - \alpha\frac{\partial h(x)}{\partial x}\right). \qquad (5.19)$$

Since this formula is inconvenient for computation, we shall simplify it. Let us examine equation

$$d\alpha h = \alpha\, dh + h\, d\alpha \qquad (5.20)$$

and estimate the order of magnitude of the terms in its right-hand part. Specific volume generally varies by approximately one unit of the fourth decimal place with a change of 10 km in the horizontal coordinate. The depth of the zero surface is on the order of 1000 m. Hence, if

$$0(\alpha) = 1, \quad 0(dh) = 10^4, \quad 0(h) = 10^5, \quad 0(d\alpha) = 10^{-4}.$$

or

$$0(\alpha\, dh) = 10^4, \quad 0(h\, d\alpha) = 10.$$

Thus, the first term in the right-hand part of equation (5.20) is by three, or by at least two to three, orders of magnitude larger than the second term. This permits us to neglect the small term and to write

$$v = \frac{1}{2\omega \sin \varphi} \left(\frac{\bar{\rho}}{\rho} \frac{\partial D}{\partial x} - \frac{\partial \alpha h}{\partial x} \right) \tag{5.21}$$

instead of (5.19).

By taking the sign of the derivative out of parentheses and placing the second term under the sign of the integral, we obtain the final expression for computing current velocity by the dynamic method relative to a "zero" surface of arbitrary complex form

$$v = \frac{1}{2\omega \sin \varphi} \frac{\partial}{\partial x} \int_{p_\mathrm{H}}^{p} (\alpha - \alpha_\mathrm{H})\, dp \tag{5.22}$$

or

$$v = \frac{D_1 - D_2}{2\omega L \sin \varphi}, \tag{5.23}$$

where

$$D = \int_{p_\mathrm{H}}^{p} (\alpha - \alpha_\mathrm{H})\, dp. \tag{5.24}$$

166 V. COMPUTING CURRENTS BY THE DYNAMIC METHOD

Here α_H is the specific volume at the zero surface. The $\dfrac{\bar{p}}{p}$ ratio is omitted because it is very close to unity.

In constructing a dynamic chart of the isobaric surface by this method, we must reduce the specific volume of

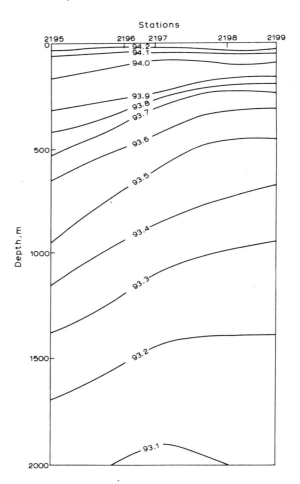

Fig. 42. Distribution of specific volume along the hydrological profile.

23. COMPUTATION OF CURRENTS

sea water by the specific volume at the "zero" surface when computing dynamic heights. No other changes are introduced into the computation scheme. Let us demonstrate it by an example.

Figure 42 represents the distribution of specific volume along a profile formed by five hydrological stations. Let the "zero" surface on the profile coincide with the $v_t = 93.2$ isostere. This agreement is not necessary in a general case.

Table 18 shows the procedure for computing the dynamic heights of isobaric surfaces above the "zero" surface at station 2196.

The dynamic heights of isobars at all other stations are computed in the same manner from the distribution of specific volume along the profile. As a result we obtain

Table 18

Computation of the dynamic heights of isobaric surfaces at station 2196; $H = 1540$ decibars

P, decibars	$v_t(p)$	$\Delta v_t = v_t(p) - v_t(p_H)$	$\overline{\Delta v_t}$	$\overline{\Delta v_t} \Delta p$	D
0	94.55	1.35	1.18	12	542
10	94.22	1.02	0.99	15	530
25	94.16	0.96	0.94	9	515
35	94.12	0.92	0.90	14	506
50	94.07	0.87			492
75	94.04	0.84	0.86	21	471
100	94.01	0.81	0.82	21	450
125	93.97	0.77	0.79	20	430
150	93.96	0.76	0.76	19	411
175	93.97	0.77	0.77	19	392
200	93.95	0.75	0.76	19	373
250	93.94	0.74	0.75	38	335
300	93.84	0.64	0.69	35	300
400	93.66	0.46	0.55	55	245
500	93.59	0.39	0.43	43	202
600	93.53	0.33	0.36	36	166
750	93.47	0.27	0.30	45	121
1000	93.38	0.18	0.23	57	64
1500	93.21	0.01	0.10	50	14
1540	93.20	0.00	0.01	14	0

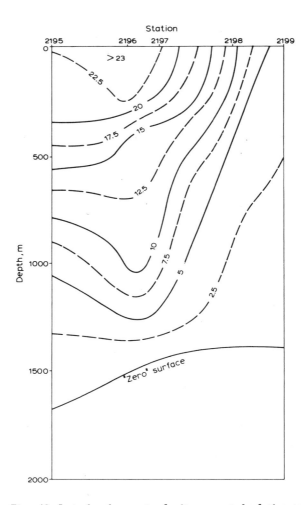

Fig. 43. Isotachs of current velocity computed relative to a "zero" surface of complex form.

the distribution of the component of the velocity of the steady gradient current normal to the profile that is shown in Fig. 43. The numbers above the isotachs give current velocity in cm/sec.

The above method for the dynamic analysis of hydrological observations along a profile or for the construction of a dynamic chart relative to a nonhorizontal "zero" surface is very simple in practice. The computation scheme differs only slightly from the computation of currents by the dynamic method relative to a horizontal reference surface. Nevertheless, we must note that positive results will be obtained by this method only if the selected nonhorizontal reference surface is in fact the "zero" or isotach surface, i.e., when the horizontal pressure gradient along it is strictly zero or is constant. Moreover, the simplification made during transition from formula (5.19) to formula (5.21) requires that the variation in water density along the zero surface be small. This condition is sufficiently well satisfied in deep ocean layers where the "zero" surface is usually located and where sea water is weakly stratified.

The above should be kept in mind in the practical application of the proposed procedure for computing currents by the dynamic method.

24. ABSOLUTE COMPUTATION OF GRADIENT CURRENT VELOCITY BY THE DYNAMIC METHOD UNDER CONDITIONS SATISFYING THE LAW OF THE PARALLELISM OF SOLENOIDS IN THE SEA

The Japanese investigators S. Haiami, H. Kawai, and M. Ouchi (1955) have proposed a procedure for computing absolute current velocities by the dynamic method when the law of the parallelism of solenoids in the sea is satisfied. Computation of currents by this method does not require preliminary determination of the depth of the no motion layer or knowledge of current velocity at some surface.

We mentioned earlier that fulfillment of the requirements of the law of the parallelism of solenoids in the sea is an indication of the presence of a gradient current that approximates the geostrophic current, except in the mixed surface layer. There is no "zero" surface in the strict sense of the word and current velocity becomes zero only at an infinitely great depth.

Let us state the Helland-Hansen and Ekman theorem of the parallelism of solenoids in the sea, since the method under consideration is based on it:

If the motion of sea water is stationary, the heat content, salinity, and amount of movement in a volume of water remain constant during motion and the water moves along an isobaric (or equipotential) surface, then the isotherms, isohalines, isotachs, and dynamic isobaths lying on the isobaric (or equipotential) surface are streamlines at this surface. If these isolines are orthogonally projected from one isobaric (or equipotential) surface on another, they will coincide with the isolines on the latter surface.

Let us now describe the method for computing current velocity. As we know, current velocity is computed in the dynamic method by the formula

$$r_i \lambda v + \operatorname{grad} D = 0,$$
$$(i = 1, 2, 3).$$
(5.25)

Here an orthogonal system of coordinates is used with the X_1 and X_2 axes lying on the isobaric surface and the X_3 axis is assumed to be directed downward along the normal to the isobaric surface. Pressure is constant at the sea surface $x_3 = 0$. In expression (5.25) r_i is the unit vector perpendicular to the unit vector of current velocity s_i and positive if it is turned clockwise relative to s_i; $\lambda = 2\omega \sin \varphi$ is the coriolis parameter; v is absolute current velocity; and D is the dynamic height of the isobaric surface.

If we designate the dynamic height of the sea surface as D_0, the dynamic height gradient of any isobaric surface can be determined from the following equation:

$$\operatorname{grad} D = \operatorname{grad} D_0 - \int_0^{x_3} \operatorname{grad} \alpha \, dx_3,$$
(5.26)

where α is the specific volume of sea water.

When the conditions of the theorem of Helland-Hansen and Ekman are satisfied, current velocity decreases continuously with depth, and therefore, by designating some positive function of the x_3, coordinate as A, we can write

$$\operatorname{grad} D = A \operatorname{grad} \alpha,$$
$$\operatorname{grad} D \neq 0, \quad \operatorname{grad} \alpha \neq 0. \tag{5.27}$$

Substituting (5.27) in (5.26), we have

$$A \operatorname{grad} \alpha = A_0 \operatorname{grad} \alpha_0 - \int_0^{x_3} \operatorname{grad} \alpha \, dx_3. \tag{5.28}$$

In the last expression A_0 and α_0 designate the values of A and α at the sea surface.

Considering that the direction of $\operatorname{grad} \alpha$ does not depend on the vertical coordinates if the law of the parallelism of solenoids in the sea is satisfied, let us differentiate (5.28) with respect to x_3. As a result we obtain

$$A \frac{d \operatorname{grad} \alpha}{dx_3} = -\left(1 + \frac{\partial A}{\partial x_3}\right) \operatorname{grad} \alpha. \tag{5.29}$$

Integration of (5.29) gives

$$\operatorname{grad} \alpha = \operatorname{grad} \alpha_0 \exp\left(-\int_0^{x_3} \frac{1 + \frac{\partial A}{\partial x_3}}{A} \, dx_3\right). \tag{5.30}$$

If we make some assumption concerning the vertical variation of function A, then formula (5.30) together with equation (5.27) will permit us to compute current velocity at depth by the dynamic method. To do this we must know the depth of the layer of no motion. The simplest assumption is that the A function is independent of x_3. Then instead of (5.30) we have

$$\operatorname{grad} \alpha = e^{-\frac{x_3}{A}} \operatorname{grad} \alpha_0. \tag{5.31}$$

This was the assumption made by the authors of the method.

To compute current velocity we must determine A (which is assumed to the constant) on the given vertical from measurements. This is done in the following manner. The specific volume gradients $[\operatorname{grad} \alpha]_1$ and $[\operatorname{grad} \alpha]_2$ at depths x_{31} and x_{32}, respectively, are computed from the distribution of the specific volume at these depths. Substituting them in (5.31), we obtain a system of two equations with two unknowns, $\operatorname{grad} \alpha_0$ and A:

$$[\operatorname{grad} \alpha]_1 = e^{-\frac{x_{31}}{A}} \operatorname{grad} \alpha_0;$$
$$[\operatorname{grad} \alpha]_2 = e^{-\frac{x_{32}}{A}} \operatorname{grad} \alpha_0. \quad (5.32)$$

It is easy to obtain A from these equations. Knowing A, let us compute from (5.27) grad D at all depths where the distribution of the specific volume is known from observations. And, finally, let us compute the sought velocity of the gradient current from formula (5.25).

The authors of the method did not specify to what conditions in the sea the assumption that function A is independent of the vertical coordinate corresponds. It is easy to demonstrate that this assumption corresponds to only one case, i.e., when the horizontal gradient of the specific volume varies exponentially along the vertical. In all other cases function A does not remain constant along the vertical.

In evaluating the foregoing method of computing the velocity of gradient currents in the sea, we must note that it can be applied only to regions where the horizontal gradient of the specific volume of water varies exponentially along the vertical. Such regions are not found everywhere in the ocean; they are usually confined to strong gradient currents. The method apparently yields good results in regions with strong currents.

CONCLUSIONS

1. If a hydrological profile, or a region under study, extends from the deep part of a basin into a shallow area where the depth is less than that of the reference surface, then in dynamic analysis that part of solid earth that lies below the selected reference surface is replaced by

24. COMPUTATION OF GRADIENT CURRENT VELOCITY 173

an imaginary water mass into which the density field is extrapolated.

2. The assumption that the horizontal pressure gradient is zero along the bottom and that the imaginary water mass is motionless is wrong. Therefore, when the density field in the imaginary water mass is extrapolated, the isopycnal surfaces along the profile must not be horizontal, but must follow their initial pattern. To eliminate subjective errors in the plotting of isopycnal lines, their slope may be kept constant at each horizon.

3. Any of the three procedures discussed here can be used to compute currents by the dynamic method relative to a nonhorizontal "zero" surface, because these procedures do not essentially differ and generally yield the same results.

4. The method of S. Haiama, M. Ouchi, and H. Kawai can be very successfully applied to regions with strong gradient currents.

CHAPTER VI

Dynamic Chart of the Sea Surface in the Kurile-Kamchatka Region of the Pacific Ocean

25. DETERMINATION OF THE "ZERO" SURFACE

In the preceding chapters we have examined general problems concerning the theory and application of the dynamic method. Let us now use some of the results obtained in an actual example and illustrate them. Let us use the material of the hydrological survey conducted in 1953 by the expeditionary ship *Vityaz'* of the Institute of Oceanology of the Academy of Sciences of the USSR. The distribution of hydrological stations is shown in Fig. 44.

The region under study extends for approximately 200 miles along the Kurile Islands and the east shore of the Kamchatka Peninsula from the southern islands of the ridge to the Kamchatka Gulf. Observations at hydrological stations were made chiefly to a depth of 1000 m, though individual stations made observations to the bottom or to a depth of several thousand meters. Deepwater stations are designated by black circles in the diagram, and 1000-m stations by light circles. The entire survey was accomplished within 73 days (May-July 1953).

During the same voyage hydrological observations were also made on a profile extending along the axis of maximum depths in the Kurile-Kamchatka basin. All stations on

25. DETERMINATION OF THE "ZERO" SURFACE

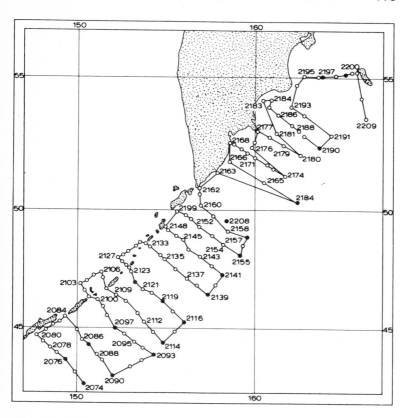

Fig. 44. Outline of the hydrological survey of the Kurile-Kamchatka region in the Pacific Ocean.

this profile reached bottom. But here we shall use only part of the observations along the longitudinal profile, because the physical-chemical properties of the water were subject to considerable seasonal variations in the two months. Moreover, the axial profile coincides with the general direction of the current in this region and is therefore less important for the study of gradient currents than the transverse profiles.

Determination of the depth of penetration of currents in this region and selection of the "zero" surface are

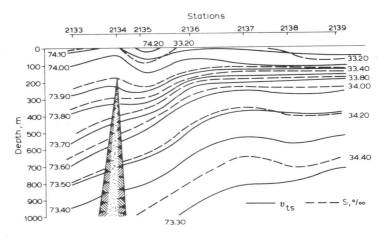

Fig. 45. Distribution of salinity (S) and of specific volume (v_{ts}) along the hydrological profile.

made difficult by the insufficient depth of observations. Most of the stations made observations to 1000 m, while deep-water stations were occupied at great distances apart, even though they were evenly distributed over the water area. This makes interpolation of deep layers difficult. The dynamic state of the water masses in this region can be analyzed only to the 1000-m level. It is impossible to construct a dynamic chart of the sea surface relative to a deeper reference surface by ordinary methods.

As we know, the Oyashio current in the Kurile-Kamchatka region has a southwesterly direction along the east coast of Kamchatka and the islands of the Kurile ridge. The peripheral waters of the Oyashio in this region differ little in their properties from the waters of the main current. Therefore there is no reason to expect the appearance here of a noticeable convective component of current velocity of advective origin. The "junction" (frontal zone) of the cold (Oyashio) and warm (Kuroshio) currents is located to the south of this region.

The structures of the temperature, salinity, and water density fields do not entirely agree in this region. The salinity and density fields generally have a similar

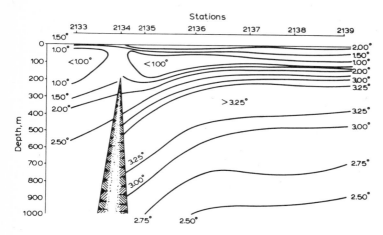

Fig. 46. Temperature distribution along the hydrological profile.

structure on vertical profiles, while the structure of the temperature field differs. But in spite of the divergence of temperature and salinity distribution on vertical profiles, the slope of both systems of isolines is generally the same. Thus, the axes of maximum and minimum temperature layers, bounded by closed isotherms on the profiles, correspond in slope to the salinity and density distribution. The isopycnal lines agree completely with the isotherms and isohalines at great depths. The vertical distribution of salinity and specific volume on the profile formed by stations 2133–2139 is shown in Fig. 45, and the temperature distribution on this profile in Fig. 46.

Analysis of other profiles, which we do not present here to save space, gives reason to believe that the requirements of the theorem on the parallelism of solenoids in the sea are generally satisfied in the Kurile-Kamchatka region of the Pacific Ocean. Hence we can draw our first important conclusions: there is apparently no strict "zero" surface in this region, current velocity decreases uniformly with depth, and the direction of the current does not change along the vertical.

Now let us examine the distribution of dissolved oxygen. We mentioned earlier that the intermediate oxygen

VI. DYNAMIC CHART OF THE SEA SURFACE

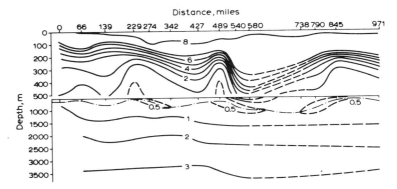

Fig. 47. Distribution of dissolved oxygen along the profile coincident with the axis of maximum depths in the Kurile-Kamchatka basin.

minimum coincides with the layer where horizontal water motion is absent or minimal. The oxygen minimum in the Kurile-Kamchatka region is located in the depth interval from 400 to 1000 m (about 0.5 ml/l).

Figure 47 shows the distribution of dissolved oxygen on the vertical profile along the axis of maximum depths of the Kurile-Kamchatka basin. The axis of the oxygen minimum on the profile is represented by a dot-dash line. As we can see from the figure, the center of the layer is located at a depth from 500 to 1000 m. The upper boundary of the intermediate oxygen minimum coincides with the intermediate maximum in the vertical stability of water layers. This testifies to the absence of a close relationship between the vertical distribution of oxygen and the horizontal motion of water. The oxygen minimum in sea water is most probably due to a combination of physical-chemical and biological factors and, to a lesser extent, to water dynamics.

Let us now analyze the structure of the salinity field in order to be able to estimate the depth of the layer of no motion by Hidaka's method. We were unable to find in the region under study a surface in the sea where the sum of the second derivatives of salinity along the horizontal coordinates $\frac{\partial^2 S}{\partial x^2} + \frac{\partial^2 S}{\partial y^2}$ is equal to zero. But it does not follow that horizontal diffusion of salt is weak or

25. DETERMINATION OF THE "ZERO" SURFACE

absent here. The vertical distribution of salinity in Kurile waters is shown in Fig. 48. The arrow indicates the depth at which the second derivative of $\frac{\partial^2 S}{\partial z^2}$ becomes zero. We can easily see that this depth can be identified with the lower boundary of the upper isohaline layer, but not with the layer of no motion. Consequently, Hidaka's method can be used only to determine the inflection point of the $S(z)$ curve, but this method does not contribute anything to the solution of the problem of the depth of current penetration.

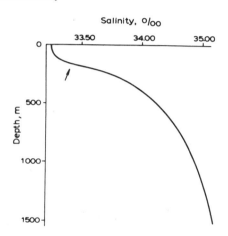

Fig. 48. Vertical distribution of salinity at hydrological station 2097.

Now let us try to get an idea of the depth of the lower boundary of the current by Parr's method, which is based on the analysis of the distortion of isopycnal layers. As we mentioned in Chapter 4, Parr suggested that the boundary between variously directed branches of circulation systems be determined from the change in the thickness of isopycnal layers, i.e., pycnomeres. He recommended that the $\frac{\Delta z}{\Delta \sigma_t}$ ratios be computed at each hydrological station and plotted on a single graph as a function of average density σ_t. According to Parr, the

VI. DYNAMIC CHART OF THE SEA SURFACE

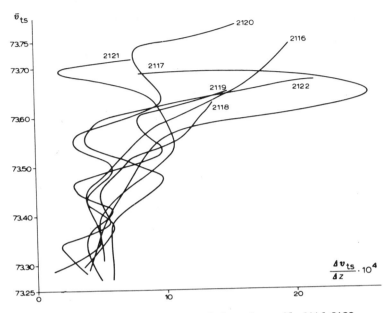

Fig. 49. Illustration of Parr's method along the profile 2116-2122.

isopycnal surface with the least distorted pycnomere must coincide with the layer of no motion or slow-moving water. But the magnitude of $\frac{\Delta z}{\Delta \sigma_t}$ ratios is too great. To avoid large values, we decided to compute $\frac{\Delta v_{ts}}{\Delta z}$ ratios. As we can see, this is sufficiently similar to Parr's characteristic. The $\frac{\Delta v_{ts}}{\Delta z}$ ratio differs from $\frac{\Delta z}{\Delta \sigma_t}$ by the multiplier, which is equal to the product of densities at two neighboring depths and is the rec procal of Parr's expression. Water density is very close to one and therefore this modification of Parr's method does not affect the results.

Figures 49 and 50 show $\frac{\Delta v_{ts}}{\Delta z}$ curves as a function of the average specific volume $\bar{v}_{ts}$ for two hydrological profiles (stations 2116-2122 and 2195-2199). The diagrams

25. DETERMINATION OF THE "ZERO" SURFACE

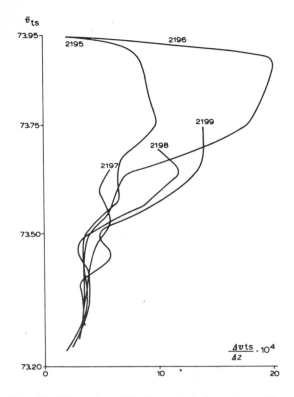

Fig. 50. Illustration of Parr's method along the profile 2195-2199.

show that the isopycnal layers generally become less distorted with depth in the upper 1000 m. This is especially clear in Fig. 50. After examining the $\frac{\Delta v_{ts}}{\Delta z}$ curves, we can see once again that the gradient component of current velocity in the region under study generally decreases with depth, since the distortion of pycnomeres decreases along the vertical. But it is very difficult to select the least distorted pycnomeres objectively. Let us assume that in Fig. 49 we have selected as such a layer the depth corresponding to the specific volume $\bar{v}_{ls} = 73.38$, because

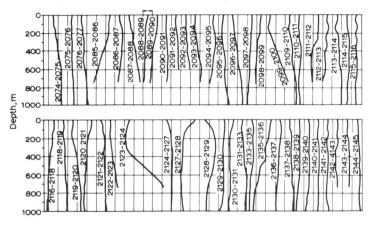

Fig. 51. Curves for the differences in the dynamic depths of isobaric surfaces between pairs of neighboring stations.

pycnomere distortions increase above and below this value. But if we consider the accuracy of the computation of specific volume ($dv_{is} = \pm\,0.02$) and the distance between neighboring levels of measurement, then some increase in the distortion of isopycnal layers with depth can be attributed to increasing computational errors, because at great depths the vertical density gradient is small and the distance between density levels is great.

In summarizing the results of the application of this method of determining the layer of no motion to conditions in the Northwest Pacific, we must note that even though Parr's method does not allow objective determination of slow-moving or motionless water, it serves as additional proof that current velocity in this region decreases with depth from the surface to a depth of 1000 m and may reach zero only at very great depths.

To estimate the depth of penetration of the gradient current in the region under study, it is very useful to analyze the differences in the dynamic depths of isobaric surfaces at many pairs of neighboring hydrological stations. The computed differences in the dynamic depths of isobaric surfaces are presented in Table 19. In examining this table we can hardly find a layer with constant

25. DETERMINATION OF THE "ZERO" SURFACE

differences, if only for the reason that there is no criterion for their constancy. It is not clear which variation of ΔD can be neglected and which value is significant. We have already mentioned that if distances between stations in the region under consideration vary and if the region extends a great distance in the meridional direction, it is difficult to find a layer where the differences in dynamic depths are uniform, because these differences are not comparable at pairs of different stations. To avoid this difficulty, the differences in the dynamic depths of isobaric surfaces were reduced to gradient current velocities at these surfaces relative to the velocity of the current at the sea surface. The results of this computation are shown in Fig. 51. All velocity curves in this graph are plotted on the same scale (each division of the horizontal scale equals 5 cm/sec). The effects of the variable distance between hydrological stations and of the change in geographic latitude are neglected here. Let us note that when the differences in the dynamic depths of isobaric surfaces were computed, the dynamic depth of the isobaric surface at the lower-numbered station was deducted from the dynamic depth of the isobaric surface at the higher-numbered station. Therefore, we can judge the direction of the component of current velocity from the data in Fig. 51 only if we consider the relative distribution of hydrological stations.

In examining the curves for the gradient component of current velocity in Fig. 51, we find that it is very difficult to ascertain the layer with relatively constant current velocity as required by Defant's method. There is apparently no such layer in the depth interval from 0-1000 m. But at the same time most of the curves exhibit a slight change in velocity with depth, with the exception of the following pairs of stations: 2085-2086, 2087-2088, 2099-2100, 2123-2124, 2135-2136, 2145-2146, 2153-2154, and 2157-2158. The main stream of the Oyashio current probably passes between these stations.

It is interesting that all curves on Fig. 51 exhibit a remarkable property. If each curve were extended downward, their curvature would decrease and velocity would approach the limiting value. This conclusion is purely qualitative. To estimate quantitatively the limiting value of

Table 19
Differences in dynamic depths, dyn·mm

z, m	2075–2074	2076–2075	2077–2076	2086–2085	2087–2086	2088–2087	2089–2088	2090–2089	2091–2090	2092–2091	2093–2092	2094–2093	2095–2094	2096–2095	2097–2096
0	0	0	0	0	0	0	0	0	0	0	0	0	0	0	0
100	9	−1	2	3	10	5	7	2	4	0	0	1	2	−5	6
200	16	−1	5	−10	13	10	−12	−13	3	6	4	6	4	−9	11
300	27	−2	9	−24	12	15	−11	−30	3	12	10	−14	3	−7	21
400	40	−3	11	−40	9	19	−12	−44	7	15	12	−22	1	−6	21
500	52	−3	13	−60	4	23	−14	−54	11	15	20	−30	3	10	29
600	60	0	15	−76	2	27	−17	−60	8	17	26	−37	8	13	37
750	64	7	20	−97	−10	32	−17	−69	4	—	—	−34	−25	19	48
1000	72	17	25	—	—	—	—	—	—	—	—	—	—	36	58

z, m	2098–2097	2099–2098	2100–2099	2110–2109	2111–2110	2112–2111	2113–2112	2114–2113	2115–2114	2116–2115	2118–2116	2119–2118	2120–2119	2121–2120	2122–2121	2123–2122
0	0	0	0	0	0	0	0	0	0	0	0	0	0	0	0	0
100	2	4	7	1	0	0	3	1	3	−10	6	1	5	5	9	2
200	6	9	1	1	2	3	2	2	14	4	−10	3	3	−17	22	2
300	7	−20	18	9	6	−17	8	4	19	4	−20	6	10	−22	23	6
400	−13	−22	36	14	14	−29	−18	11	19	4	−28	9	23	−24	19	14
500	−15	−26	54	16	15	−35	−25	17	18	3	−34	11	32	−25	20	21
600	−13	−26	72	23	7	−35	−23	18	18	3	−40	14	40	−30	23	29
750	−26	—	98	34	4	−32	−34	36	6	6	−46	17	51	−38	24	46
1000	−35	—	—	36	6	—	—	38	6	6	−53	16	59	−41	31	—

25. DETERMINATION OF THE "ZERO" SURFACE

z, m	2124−2123	2127−2124	2128−2127	2129−2128	2130−2129	2131−2130	2133−2131	2135−2133	2136−2135	2137−2136	2138−2137	2139−2138	2140−2139	2141−2140
0	0	0	0	0	0	0	0	0	0	0	0	0	0	0
100	6	26	−30	24	0	−7	7	8	−9	4	3	−1	−1	4
200	9	31	−36	32	−1	−12	0	12	−19	2	5	0	−1	−7
300	6	35	−39	34	−2	−11	5	13	−45	6	6	0	−1	0
400	8	37	−40	31	−2	−12	10	8	−32	−10	8	0	−1	0
500	26	38	−43	27	−3	−13	14	−2	−83	−15	10	3	−2	0
600	50	34	−45	21	−5	−10	16	−16	−94	−22	14	7	−1	0
750	71	41	−44	8	−5	−10	17	−31	−112	−29	19	−14	9	−6
1000	−	33	−43	4	−14	−14	14	−38	−135	−36	24	−34	25	−12

Station

z, m	2142−2141	2143−2142	2144−2143	2145−2144	2146−2145	2147−2146	2151−2147	2152−2151	2153−2152	2154−2153	2155−2154	2156−2155	2157−2156	2158−2157	2159−2158	2160−2159
0	0	0	0	0	0	0	0	0	0	0	0	0	0	0	0	0
100	1	−1	0	1	0	0	11	3	−5	−7	−38	37	3	−1	−1	−12
200	0	−3	5	7	−3	6	14	2	−5	−15	−39	35	1	−7	−7	−10
300	3	−12	9	13	5	14	23	0	−6	−32	−46	36	2	−28	−12	−12
400	1	−10	12	15	13	13	48	5	−11	−53	−51	35	4	−52	−20	−12
500	1	−10	16	13	23	−1	72	10	−16	−72	−57	33	6	−72	−32	−14
600	2	−9	17	14	31	7	79	11	−17	−90	−62	33	8	−90	−43	−13
750	2	−2	16	15	42	28	73	13	−16	−114	−72	36	14	−111	−58	−7
1000	5	−	−	15	−	−	−	−	−16	−144	−46	14	11	−	−	−

Station

current velocity and the depth at which this value is reached requires data on the temperature and salinity of deep layers. Unfortunately we possess few such data. This circumstance makes it impossible to estimate strictly the depth of the lower boundary of the gradient current by Defant's method. Application of this method shows that in the region under study current velocity decreases with depth, and apparently becomes insignificant everywhere at a depth on the order of 2500-3000 m. Velocity reaches small values more rapidly where surface currents are weak. In the core of the Oyashio the current penetrates to a greater depth, but gradient current velocity must be small enough everywhere at the 3000-m level and deeper. Strictly speaking, using Defant's method we were able to find the depth at which the velocity of the gradient current does not vary significantly along the vertical, but is not necessarily small in absolute magnitude. We shall discuss this in more detail below.

Now let us try to determine the vertical distribution of the gradient component of current velocity by Hidaka's method, which is based on the dynamic method and assumes that the water volume and the mass of salt are continuous in vertical prisms formed by triangles of hydrological stations. In Chapter 4 we came to the conclusion that it is impossible to obtain the true picture of the vertical distribution of current velocity by this method, given the existing accuracy of initial measured data (water temperature and salinity).

Let us now investigate the applicability of Hidaka's method to other conditions. Let us take four hydrological stations: 2074, 2076, 2087, and 2090. Depths at these stations vary from 4000 to 8000 m and observations on temperature and salinity distribution were made to the bottom.

In our opinion, replacement of depths at hydrological stations by average values in the intervals between pairs of stations is unacceptable in this case, because the volume of water in the prisms would have a complex lower boundary that would not agree with the bottom relief which is extremely complex in this region. Therefore let us limit the prism at the bottom by a horizontal surface $z = 4000$ m, assuming that water movement is insignificant below. This assumption is not unnatural because the depth at one of the four stations is 4000 m, and current velocity must necessarily be low in a deep-water basin. Moreover, this would not affect the results of the computation. We shall demonstrate below the incompatibility of the system of equations.

The relative location of hydrological stations is shown in Fig. 52. The arrows indicate directions that are assumed to be positive on the corresponding faces of the prism. The figures are the successive numbers of the faces. The distances between hydrological stations are:

$L_1 =$ 74.6 miles $= 286 \cdot 10^{-4}$ conventional units
$L_2 =$ 82.2 ,, $= 315 \cdot 10^{-4}$,, ,,
$L_3 =$ 73.0 ,, $= 280 \cdot 10^{-4}$,, ,,
$L_4 =$ 99.1 ,, $= 380 \cdot 10^{-4}$,, ,,
$L_5 = 127.5$,, $= 489 \cdot 10^{-4}$,, ,,
$L_6 = 104.4$,, $= 400 \cdot 10^{-4}$,, ,,

Henceforth we shall use conventional units of measurement of distances L_i. This will not affect the result, but will simplify computation considerably.

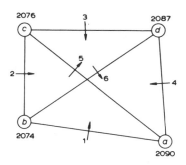

Fig. 52. Arrangement of four hydrological stations for computation of the vertical structure of the gradient current by Hidaka's method.

The vertical distribution of initial values (salinity and the dynamic depths of isobaric surfaces) at the four hydrological stations is given in Table 20. Table 21 gives the vertical distribution of current velocity in the intervals between stations relative to current velocity at the sea surface. The gradient components of current velocity are computed from the distribution of water density by the ordinary dynamic method.

To compute coefficients entering the system of equations (4.11) and given by the expressions in (4.12), we used numerical integration of functions—the trapezoidal method. As a result, the following values for the functions were obtained:

$A_1 = 114.4;$ $H_1 = 3938.45;$ $M_1 = -1009.41;$ $N_1 = -33486.98;$
$A_2 = 126.0;$ $H_2 = 4351.48;$ $M_2 = 1162.41;$ $N_2 = 38553.01;$
$A_3 = 112.0;$ $H_3 = 3862.14;$ $M_3 = 332.14;$ $N_3 = 11126.76;$
$A_4 = 152.0;$ $H_4 = 5242.55;$ $M_4 = -736.14;$ $N_4 = -24519.17;$
$A_5 = 195.6;$ $H_5 = 6762.41;$ $M_5 = 354.87;$ $N_5 = 12005.12;$
$A_6 = 160.0;$ $H_6 = 5523.39;$ $M_6 = 1174.64;$ $N_6 = 48906.51.$

To determine current velocity c_i at the sea surface, we obtain the system of equations

$$\left.\begin{array}{l} 114.4c_1 + 126.0c_2 - 195.6c_5 - 201.9 = 0, \\ 112.0c_3 + 152.0c_4 + 195.6c_5 - 149.1 = 0, \\ 114.4c_1 + 152.0c_4 + 160.0c_6 - 570.9 = 0, \\ 114.0c_1 + 126.0c_2 - 195.8c_5 - 200.9 = 0, \\ 112.0c_3 + 152.0c_4 + 196.1c_5 - 40.2 = 0, \\ 114.1c_1 + 151.9c_4 + 160.0c_6 - 263.6 = 0. \end{array}\right\} \quad (6.1)$$

We reduced the fourth equation by 34.53, the fifth equation by 34.49, and the sixth by 34.52 to obtain more comparable results. These values are equal to the average salinity at the lateral faces of the trihedral prisms.

We can see from the system of equations (6.1) that with the unknowns in the three last equations the coefficients differ from the coefficients with the corresponding unknowns in the first three equations by values that do not exceed the limits of accuracy of the computations:

25. DETERMINATION OF THE "ZERO" SURFACE

$A_1 - H_1 = 0.4;\ 0.3;$ $A_4 - H_4 = 0.0;\ 0.1;$
$A_2 - H_2 = 0.0;$ $A_5 - H_5 = -0.2;\ -0.5;$
$A_3 - H_3 = 0.0;$ $A_6 - H_6 = 0.0.$

Table 20

Station

z, m	2076		2087		2090		2074	
	S, ‰	D dyn·mm	S, ‰	D dyn·mm	S, ‰	D dyn·mm	S, ‰	D dyn·mm
0	33.28	0	33.21	0	33.17	0	33.31	0
10	33.27	41	33.24	42	33.18	42	33.31	40
25	33.28	103	33.24	104	33.22	104	33.31	101
50	33.30	205	33.24	207	33.26	205	33.34	201
100	33.30	406	33.29	408	33.30	404	33.37	398
150	33.41	603	33.47	605	33.70	598	33.46	591
200	33.64	796	33.68	798	33.96	783	33.86	781
250	33.81	985	33.83	986	34.05	965	33.99	965
300	33.90	1170	33.91	1171	34.11	1145	34.11	1145
400	34.03	1534	34.03	1534	34.22	1497	34.20	1497
500	34.14	1890	34.14	1888	34.25	1843	34.28	1841
600	34.22	2240	34.21	2237	34.25	2188	34.33	2181
750	34.33	2753	34.33	2752	34.35	2698	34.43	2682
1000	34.45	3585	34.42	3588	34.46	3523	34.55	3494
1500	34.59	5193	34.55	5213	34.57	5131	34.65	5069
2000	34.67	6753	34.61	6798	34.64	6703	34.70	6609
2500	34.70	8293	34.66	8360	34.68	8248	34.72	8129
3000	34.70	9823	34.68	9902	34.70	9783	34.78	9629
4000	34.72	12893	34.69	12982	34.71	12853	34.80	12619

Some differences have two values here, because coefficients with subscripts 1, 4, and 5 enter two equations and are divided by different divisors upon reduction.

Let us examine the accuracy with which coefficients A_i and H_i are computed. It follows directly from (4.12) that

$$\frac{dA_i}{A_i} = \frac{dL_i}{L_i} + \frac{dh_i}{h_i}. \qquad (6.2)$$

Let us apply the mean value theorem to the expression for H_i, then

$$H_i = L_i h_i \overline{S}_i, \qquad (6.3)$$

Table 21

Current velocity, cm/sec

z, m	Station					
	2074–2090	2074–2076	2076–2087	2087–2090	2090–2076	2074–2087
0	0	0	0	0	0	0
10	− 0.14	0.06	0.07	0	−0.04	0.10
25	− 0.21	0.13	0.07	0	−0.04	0.15
50	− 0.28	0.26	0.14	−0.11	0	0.30
100	− 0.42	0.52	0.14	−0.22	0.08	0.50
150	− 0.49	0.78	0.14	−0.38	0.20	0.70
200	− 0.14	0.98	0.14	−0.82	0.52	0.85
250	0	1.30	0.07	−1.16	0.80	1.05
300	0	1.65	0.07	−1.43	1.00	1.30
400	0	2.40	0	−2.04	1.48	1.85
500	− 0.14	3.18	−0.14	−2.48	1.88	2.35
600	− 0.49	3.96	−0.21	−2.70	2.08	2.80
750	− 1.12	4.61	−0.07	−2.97	2.20	3.50
1000	− 2.03	5.91	0.21	−3.58	2.48	4.70
1500	− 4.34	8.08	1.40	−4.51	2.48	7.10
2000	− 6.58	9.36	3.15	−5.22	2.00	9.45
2500	− 8.34	10.67	4.69	−6.16	1.80	11.55
3000	−17.78	12.60	5.53	−6.55	1.60	13.65
4000	−23.41	17.80	6.09	−7.10	1.60	13.35

where $\overline{S}_i$ is the mean value of salinity at the face of the prism.

The relative mistake in the computation of coefficient H_i is estimated from the expression

$$\frac{dH_i}{H_i} = \frac{dL_i}{L_i} + \frac{dh_i}{h_i} + \frac{d\overline{S}_i}{\overline{S}_i}. \tag{6.4}$$

As we established earlier, the relative mistake in the determination of sea water salinity from chlorine is $\pm 0.009\%$. The distance between hydrological stations can be determined with an error of the order of ± 0.2 miles and sometimes less accurately. In this case

$$\frac{dL_i}{L_i} \approx \pm 0.3\%;$$

25. DETERMINATION OF THE "ZERO" SURFACE

$$\frac{dh_i}{h_i} \approx \pm 1.0\%;$$

$$\frac{dS_i}{\overline{S_i}} \approx \pm 0.2\%.$$

Consequently, the actual error in the computation of coefficients A and H_i exceeds $\pm 1\%$. At the same time, the $A_i - H_i$ difference is only a small fraction of a percent of the absolute values of any of these coefficients, or even zero. In other words, the difference between the A_i and H_i coefficients is entirely within the limits of probable computational errors. All this shows clearly that the determinant of the system (6.1) is zero and the system is incompatible. We arrived at the same results when examining the example calculated by the author of the method (Chapter 4). Consequently, we have shown once more that Hidaka's method is unsuitable for the determination of the layer of no motion in the sea, because the system of equations cannot be solved.

The following conclusions can be drawn from the application of various methods to the evaluation of the depth of penetration of gradient currents in the region under consideration, and on the basis of some general considerations about the vertical structure of the steady gradient current: a) The velocity of the gradient current in this region decreases regularly with depth; b) the direction of the gradient component of current velocity does not change noticeably along the vertical; c) there is apparently no oppositely directed gradient current in deep layers; and d) the lower boundary of the current (if the current does not penetrate at high velocity to the bottom) is located below 1000 m, presumably at 2500-3000 m. If we consider the lower boundary of the current to be the isoline of the low modulus of current velocity, it cannot be horizontal in this region. This surface sinks in regions with strong surface currents, but is closer to the surface where current velocity at the sea surface is low. If we consider the lower boundary of the current to be the surface at which current velocity comprises a given small fraction of the velocity of the gradient current at the surface, then this boundary deflects insignificantly from the horizontal.

The following can be said concerning the determination of the "zero" surface in the Kurile-Kamchatka region. Regardless of the method used, analysis of the structure of the sea water density field gives only the pattern of the vertical variation of the gradient component of the velocity of a steady current, but does not make it possible to judge the absolute values of gradient current velocity. To obtain absolute values we must take into consideration the force that induces the current or, for a rough evaluation, any additional argument. Even if we were able to determine the depth below which current velocity remains almost constant along the vertical, it is no proof of insignificant velocity in this layer. It could prove that even though current velocity does not vary in a deep layer, it is high there. Consequently, the relative magnitude of the gradient component of velocity of a steady current can be computed from the structure of the density field. Absolute computation is possible only if the forces that induce the system of currents in the sea are taken into account. Data on the structure of sediment in a given region may prove to be very useful in some cases. If a current penetrates to the bottom at high velocity, large vertical velocity gradients arise (as a result of adhesion at the solid boundary in the bottom layer) that promote erosion of sediments and the enlargement of sediment particles at the bottom. Fine silty sediments must be found in those regions where the velocity of bottom currents is low.

To determine the depth of the surface below which the gradient component of the velocity of a steady current does not change noticeably along the vertical, i.e., below which the density field has little effect on the gradient current, stratification functions $\Phi(z)$ and $F(H)$ were computed with various artificially assigned reference surface depths. The results of the computations are presented in Tables 22 and 23.

Let us recall that the expressions for the stratification function were obtained on the assumption that the density distribution of sea water corresponds to Shtokman's model. Direct verification of the applicability of the model showed that it is not fully satisfied in the region under consideration. The closest values for the influence function

Table 22

Function $\Phi(z)\cdot 10^2$ with $H = 3000$ m

$H-z$, m	\multicolumn{10}{c}{Station}									
	2074	2076	2087	2090	2114	2116	2119	2139	2141	2157	2208
0	100	100	100	100	100	100	100	100	100	100	100
50	92	92	91	92	92	92	92	91	91	92	94
100	85	86	85	85	85	84	84	82	83	86	89
200	73	73	74	72	73	71	71	69	69	74	79
300	63	63	64	63	64	60	62	59	58	65	70
400	55	54	56	56	56	52	54	50	50	57	62
500	48	47	48	49	49	45	47	43	42	50	55
600	41	40	42	42	43	39	40	36	36	44	
750	33	31	34	34	35	31	32	29	28	36	40
1000	28	21	24	23	24	21	22	20	19	24	28
1500	15	8	11	10	10	9	10	8	6	10	14
2000	7	2	4	3	3	4	4	2	1	3	6
2500	2	0	1	1	1	1	1	0	0	1	1
3000	0	0	0	0	0	0	0	0	0	0	0

Table 23

Stratification function $F(H)\cdot 10^5$, cm^{-1}

H, m	\multicolumn{12}{c}{Station}											
	2074	2076	2087	2093	2093	2114	2116	2119	2139	2141	2157	2164	2208
1500	2.60	2.42	2.48	2.49	2.51	2.45	2.57	2.69	2.64	3.04	2.35	2.45	2.24
2000	2.14	2.00	2.06	1.97	1.92	1.88	2.09	2.09	1.99	2.04	1.88	2.07	1.82
2500	1.79	1.70	1.66	1.69	1.72	1.74	1.93	1.72	1.74	1.82	1.62	1.70	1.59
3000	1.39	1.63	1.48	1.49	1.51	1.52	1.61	1.56	1.74	1.82	1.50	1.62	1.33
4000	—	1.41	1.39	—	1.34	1.20	1.30	1.33	—	—	1.32	1.24	1.28
5000	—	1.33	—	—	—	—	—	—	—	—	—	—	1.24	1.20
6000	—	1.33	—	—	—	—	—	—	—	—	—	—	—	1.22

$f(x, y)$ on the vertical were obtained in the region of strong currents, but even here function $f(x, y)$ cannot be considered as being independent of the vertical coordinate. However, we found it possible to use the method of estimating the depth of penetration of gradient currents based on the density model, because when the accuracy of

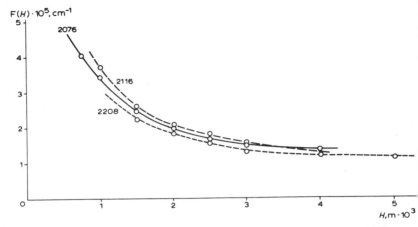

Fig. 53. Graphical representation of the stratification function $F(H)$ computed from the data for three hydrological stations.

density determinations is considered in the verification of the model, all deviations are within the limits of probable computational error. We considered the small variation of the stratification function along the horizontal as proof of the similarity of the vertical structure of water masses. As an illustration, Fig. 53 gives curves of the behavior of function $F(H)$, computed from observational data at hydrological stations 2076, 2116, and 2208. These stations are located in different parts of the region under study. From Fig. 53 and from the estimate of the influence function, it follows that the density field in this region corresponds in its most general features to Shtokman's model, but the details of its structure do not fit the model (possibly because of measurement errors).

Sea water stratification has the strongest influence on the gradient current in the upper layer; the vertical velocity gradient decreases with depth, and becomes practically negligible starting from a depth of 2500-3000 m. This is clearly shown in Fig. 53. Consequently, the 3000-decibar reference surface can be used to construct a dynamic chart of the surface of the Kurile-Kamchatka region of the Pacific. For the reasons mentioned above, such a chart will not represent the absolute relief of the sea surface, but will only characterize the velocity of the steady gradient current at the surface relative to current velocity at the 3000-m or any other underlying level, since current velocity remains practically constant along the vertical in deep water.

In the region under study, sediment at a depth of several thousand meters consists of fine silt particles; coarser sediment is found only in shallow water. This indicates that at great depths, in particular at the 3000-m level, current velocity is sufficiently small.

Let us consider the 3000-decibar isobaric surface as the "zero" surface in the computation of current velocity at the sea surface by the dynamic method.

26. CONSTRUCTION OF A DYNAMIC SEA SURFACE CHART

We have established that the velocity of the steady gradient current is low at a depth of 3000 m and selected this surface as the "zero" surface for the construction of a dynamic chart of the sea surface. But there are very few hydrological stations in the region under study where observations were made to such depth. Measurements at most of the stations were made to a depth of 1000 m. Therefore, let us first construct a relief chart of the sea surface relative to a 1000-decibar reference surface, and then reduce the gradient current velocities obtained to absolute values, using the relationships obtained earlier on the basis of Shtokman's density model.

The dynamic chart of the sea surface relative to the 1000-decibar surface is represented in Fig. 54. The dynamic contours are plotted at intervals of 20 dyn·mm. This chart clearly shows the current flowing to the southwest along the Kamchatka coast and the islands of the Kurile ridge. Several eddies of varying intensity are located along the left edge of this current. Current velocity in the eddies is relatively small.

In examining the accuracy of the dynamic method of computing sea currents we showed that, owing to the accumulation of random errors, the computed dynamic

VI. DYNAMIC CHART OF THE SEA SURFACE

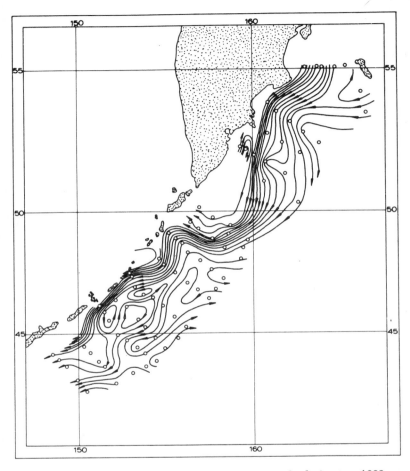

Fig. 54. Dynamic relief of the sea surface computed relative to a 1000-decibar reference surface.

height differs from the actual dynamic height of the isobaric surface by a value that can be estimated, knowing the method of measurement of the initial data. In our case, the maximum error in the computation of the dynamic height of the sea surface was ± 20 dyn·mm with $H = 1000$m. Water temperature was measured with deep-water

26. CONSTRUCTION OF SEA SURFACE CHART

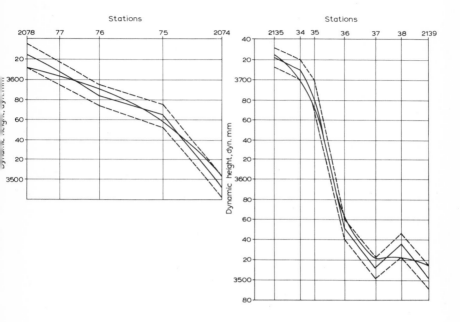

Fig. 55. Dynamic profile of the sea surface along two sections, computed relative to a 1000-decibar reference surface.

reversing thermometers with scale divisions of $0.1°C$, and water salinity was determined by the standard chemical method from chlorine with buret scale divisions of 0.01 ml. Considering that an error in dynamic height that is equal to half its maximum value is real, it becomes possible to smooth the dynamic relief of the sea surface within ±10 dyn·mm. Smoothing is not unnatural, since we are sure that the accuracy of dynamic height does not exceed ±10 dyn·mm.

Figure 55 shows sections of the dynamic topography of the sea surface along two hydrological profiles. Similar sections were constructed from the data of all other hydrological profiles. The thin broken line in the graph represents the computed dynamic profile and the dashed

198   VI. DYNAMIC CHART OF THE SEA SURFACE

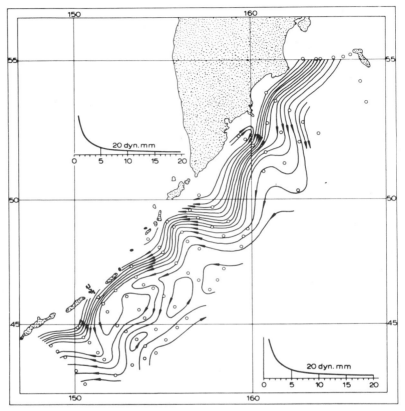

Fig. 56. Smoothed dynamic relief of the sea surface computed relative to a 1000-decibar reference surface.

lines are the boundaries of the interval of errors that are removed by ± 10 dyn·mm from the computed profile. The thick line in the interval of errors represents the smoothed dynamic profile of the sea surface. In plotting this line we tried to make it as smooth as possible, because any distortion in it within the interval of errors is unreliable.

Now we must merely determine the new dynamic height values for each station from the smoothed curve in order to construct a dynamic chart. We did it differently,

26. CONSTRUCTION OF SEA SURFACE CHART

using dynamic profiles for graphical interpolation in the intervals between hydrological stations. As a result we obtained a chart of the smoothed dynamic relief of the sea surface with a reference surface at a depth of 1000 m. This chart is shown in Fig. 56. The isolines on the chart are plotted every 20 dyn·mm and the nomograms correspond to latitudes $45°$ and $55°$.

The main features of the computed sea surface relief are retained in the chart of the smoothed relief. As before, there is a rather strong southwesterly current, but part of the small eddies are absent on this chart, because they were within the limits of accuracy of the computation.

Using the results obtained in Chapter 3, let us reduce relative current velocities to absolute. By absolute velocity of the steady gradient current at the sea surface we mean current velocity computed by the dynamic method relative to the 3000-decibar surface. The "zero" surface is located at a depth of 3000 m in this region.

The stratification function $\Phi(z)$ characterizes the ratio of the gradient component of current velocity at depth z to the velocity of the gradient current at the sea surface. From Table 22 it follows that with $H = 3000$ m, the velocity of the gradient current at a depth of 1000 m averages 25% of the velocity of the current at the sea surface. Consequently, to obtain absolute velocity values we must multiply current velocity obtained from the dynamic chart (Fig. 56) by a coefficient of 1.33. As a result we obtained the scheme of gradient currents in the Kurile-Kamchatka basin that is shown in Fig. 57. The numbers between arrows give current velocity in cm/sec. It should be noted that in computing current velocities our aim was to obtain values that would characterize as large an area as possible around a particular point.

The computed and measured current velocities at the sea surface agree rather well. There is some deviation that is attributable to the fact that a dynamic chart always gives a smoothed picture of circulation that is averaged over the horizontal. The extremes in the relief of an isobaric surface are usually inaccurate on a dynamic chart, because each pair of hydrological stations gives the average transport of water between stations.

200   VI. DYNAMIC CHART OF THE SEA SURFACE

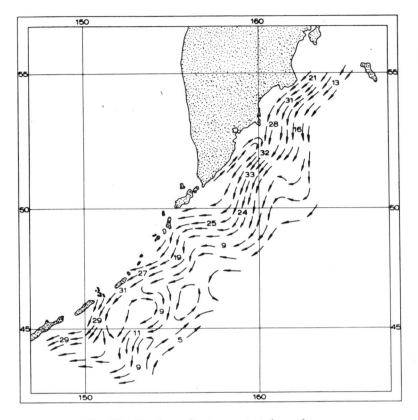

Fig. 57. Steady gradient current at the surface.
The numbers near the arrows give current velocity in cm/sec.

We shall not examine the pattern of currents in the Kurile-Kamchatka region of the Pacific Ocean in more detail, nor explain its characteristics. The purpose of this chapter, i.e., to illustrate the results obtained and give a general idea of their use in practical computations of gradient currents by the dynamic method, has been accomplished. We have determined the depth below which the velocity of steady gradient currents is negligibly small, have computed current velocities at the sea surface, and have obtained generally plausible results. The analysis performed in this chapter can be used in the computation of gradient currents in any ocean region and any sea.

CONCLUSIONS

1. Use of various methods of determining the depth of the "zero" surface in the Kurile-Kamchatka region of the Pacific did not yield positive results, with the exception of Defant's and Parr's methods. Analysis of the differences in the dynamic depths of isobaric surfaces and the distortion of the thickness of isopycnal layers showed that in this region the velocity of the gradient current generally decreases with depth. Significant conclusions could not be drawn because of limited observations in the deep layers.

2. The depth below which the density field has no significant effect on the gradient component of current velocity has been found from the integral stratification function $F(H)$. It is on the order of 3000 m.

3. Analysis of water stratification has shown that the velocity of the gradient current at the depth of the reference surface ($z = 1000$ m) is approximately 25% of the velocity of the current at the sea surface. This made it possible to introduce a correction factor into computed velocities.

4. Evaluation of the accuracy of computation of dynamic heights shows that a number of eddies indicated at the periphery of the main current cannot be reliably determined. These eddies cannot be contours on the chart of the smoothed dynamic relief.

5. The methods used in this chapter to estimate the depth of penetration of the gradient current and the method of constructing a dynamic chart as described here can be used in the computation of currents by the dynamic method in any region of the World Ocean.

RESUME

The dynamic method of computing sea currents that was developed more than half a century ago remains to this day the only method of obtaining data on water circulation in many parts of the World Ocean. Theoretical studies of sea currents have disclosed the major causes of water circulation in oceans and the pattern of the vertical distribution of current velocities in an inhomogeneous sea, and have elucidated the relationship between the velocity field and density distribution. We now understand the advantages and shortcomings of the dynamic method. But many problems having a direct bearing on current velocity computations from density distribution in the sea require detailed study.

The assumption that gradient currents approximate the geostrophic currents must be verified theoretically and experimentally. Until now, comparisons of the results of computation by the dynamic method with the results of instrumental measurements have been qualitative in nature. Moreover, such comparisons have been possible only for the more readily measured surface layers. The influence of friction on the structure of the gradient current has not been sufficiently studied. Also of great importance is the study of the rates of readjustment of the density field during the mutual adjustment of the distribution of velocities and density in the sea. The fact is that the observed density distribution is not stationary, but contains short-period disturbances that may not be associated with the averaged water circulation.

It is quite obvious that we did not attempt, and could not attempt, to examine the above problems. The main purpose of our work was to give a correct interpretation of the dynamic method on the basis of accumulated information and data, to examine the most important aspects

of its application, and to correct those inaccuracies
that can still be encountered in the literature devoted to
the dynamic method. Greatest attention was paid to methods
of determining the "zero" surface, as it is the basic
problem in the practical use of the dynamic method.
We must note that the unsuccessful attempts of many investigators to develop methods for the indirect determination of the layer of no motion in the sea are not accidental.
The structure of the density field, or the distribution of
other oceanological elements in the sea, determines the
intensity of the vertical change in the velocity of the gradient current and is not associated with its absolute value.
Forces that induce motion must be considered in order to
obtain true velocity values.

It is to be hoped that in the near future the combination
of theoretical investigations and experimental studies
under actual conditions will widen and deepen our knowledge
concerning the dynamics of sea currents, and will help
create new and more perfect computational methods.

Bibliography

BJERKNES, V., 1900. Das dynamische Prinzip der Zirkulationsbewegung in der Atmosphäre. Meteorol. Z.
BJERKNES, V. and SANDSTRÖM, I., 1912. Dynamische Meteorologie und Hydrographie, v. 1.
BOEHNECKE, G., 1932. Das ozeanographische Beobachtungsmaterial. Wiss. Ergeb. "Meteor" 1925-1927, Vol. 4, Part 2.
DANOIS, Ed., 1934. Les transgressions océaniques. Off. Sci. Tech. Pêch. Maritimes, Rev. Trav., v. 7, no. 28.
DEFANT, A., 1941a. Die absolute Topographie des physikalischen Meeresniveaus und der Druckflächen, sowie die Wasserbewegungen im Atlantischen Ozean. Deutsche Atlantische Exped. "Meteor" 1925-1927, Vol. 6, Part 2, no. 5.
DEFANT, A., 1941b. Die absolute Berechnung ozeanischer Ströme nach dem dynamischen Verfahren. Ann. Hydrograph. Marit. Meteorol., no. 6.
DEFANT, A., 1954. Turbulenz und Vermischung im Meere. Deut. Hydrograph. Z., v. 7, no. 1/2.
DIETRICH, G., 1936. Aufbau und Bewegung von Golfstrom und Agulhasstrom. Naturwissenschaften, no. 15.
DIETRICH, G., 1937. Die Lage der Meeresoberfläche im Druckfeld von Ozean und Atmosphäre. Veröffentl. Inst. Meeresk., New Ser., A, v. 33.
DOBROVOL'SKII, A. D., 1949. Contribution to the problem of the location of the zero surface for dynamic computation in the North Pacific. Tr. Inst. Okeanol. Akad. Nauk SSSR, v. 4.
DROGAITSEV, D. A., 1954. Structure of the wind field over the sea. Tr. Inst. Okeanol. Akad. Nauk SSSR, v. 9.
EKMAN, V. W., 1905. On the influence of the earth's rotation on ocean currents. Arkiv Mat., Astron. Fysik, v. 2, no. 11.

EKMAN, V. W., 1923. Über Horizontalzirkulation bei winderzeugten Meeresströmungen. Arkiv Mat., Astron. Fysik, v. 17, No. 26.
EKMAN, V. W., 1926. Können Verdunstung und Niederschlag im Meere merkliche Kompensationsströme verursachen? Ann. Hydrograph. Marit. Meteorol., no. 4.
EXNER, F. M., 1912. Zur Kenntnis der unterstehenden Winde über Land und Wasser und der durch sie erzeugten Meeresströmungen. Ann. Hydrograph. Marit. Meteorol., no. 4.
FEDOROV, K. N., 1956. Results of the simulation of full currents induced by the wind in the sea. Tr. Inst. Okeanol. Akad. Nauk SSSR, v. 19.
GOLDSBROUGH, G. R., 1933. Ocean currents produced by evaporation and precipitation. Proc. Roy. Soc., Ser. A, v. 141, no. 845.
GROEN, P., 1948. Methods for estimating dynamic slopes and currents in shallow water. J. Marine Res., v. 7, no. 3.
HAIAMI, S., KAWAI, H., and OUCHI, M., 1955. On the theorem of Helland-Hansen and Ekman and some of its applications. Rec. of Oceanogr. Works in Japan, v. 2, no. 2.
HELLAND-HANSEN, B., 1934. The Sognefjord section. James Johnstone Memorial Volume, Liverpool.
HIDAKA, K., 1940a. Absolute evaluation of ocean currents in dynamical calculations. Proc. Imp. Acad., Tokyo, v. 16.
HIDAKA, K., 1940b. A practical evaluation of ocean currents. Proc. Imp. Acad., Tokyo, v. 16.
HIDAKA, K., 1949. Depth of motionless layer as inferred from the distribution of salinity in the ocean. Trans. Am. Geophys. Union, v. 30, no. 3.
HIDAKA, K., 1950. An attempt to determine the motionless layers in the ocean. Oceanogr. Mag., v. 2, no. 1.
ISELIN, C. O'D., 1936. A study of the circulation of the western North Atlantic. Papers Phys. Oceanogr. Meteorol., v. 4, no. 4.
JACOBSEN, J.P., 1916. Contribution to the hydrography of the Atlantic. Medd. Komm. Havundersgelser, Ser. Hydrografi, v. 2.

JACOBSEN, J. P. and JENSEN, A. J., 1926. Examination of hydrographical measurements from the research vessels "Explorer" and "Dana" during the summer of 1924. Rapp. Conseil Perm. Intern. Exploration Mer, v. 39.

JACOBSEN, J. P., 1929. Contribution to the hydrography of the North Atlantic. Danish "Dana" Expedition 1920-1922. Oceanogr. Repts., no. 3.

JAKHELLN, A., 1936. The water transport of the gradient currents. Geofys. Publikajoner, v. 11, no. 11.

KITKIN, P. A., 1953. Circulation induced by wind in a sea of variable density. Izv. Akad. Nauk SSSR, Ser. Geofiz., no. 3.

LAIKHTMAN, D. L. and CHUDNOVSKII, A. F., 1949. Physics of the air layer near the ground. Leningrad-Moscow.

LINEIKIN, P. S., 1955. Wind-driven currents in the baroclinic layer of the sea. Tr. Gos. Okeanogr. Inst., no. 29(41).

MAMAEV, O. I., 1955. Methods of determining the zero dynamic surface in the World Ocean. Vestn. Mosk. Univ., no. 10.

MOHN, H., 1885. Die Strömungen des Europäischen Nordmeeres. Petermanns Geogr. Mitt., Ergänzungsheft, no. 79.

NEUMANN, G., 1942. Die absolute Topographie des physikalischen Meeresniveaus und Oberflächenströmungen des Schwarzen Meeres. Ann. Hydrograph. Marit. Meteorol., v. 70, no. 9.

NEUMANN, G., 1955. On the dynamics of wind-driven current. Meteorol. Papers, v. 2, no. 4.

PARR, A., 1938. Analysis of current profiles by a study of pycnomeric distortion and identifying properties. J. Marine Res., no. 4.

POST, L. A., 1954. A practical method for the prediction of the surface currents of the ocean. Tellus, v. 6, no. 1.

PRANDTL, L., 1951. Hydroaeromechanics. Moscow.

REID, R. O., 1948. A model of the vertical structure of mass in equatorial wind-driven current of a baroclinic ocean. J. Marine Res., v. 7, no. 3.

RILEY, G. A., 1951. Oxygen, phosphate and nitrate in the Atlantic Ocean. Bull. Bingham Oceanogr. Coll., v. 13.

SANDSTRÖM, I. W. and HELLAND-HANSEN, B., 1903. Über die Berechnung von Meeresströmungen. Rept. Norw. Fish. and Mar. Inst.

SARKISYAN, A. S., 1954. Computation of stationary wind-driven currents in the ocean. Izv. Akad. Nauk SSSR, Ser. Geofiz., no. 6.

SEIWELL, H. R., 1934. The distribution of oxygen in the western basin of the North Atlantic. Papers Phys. Oceanogr. Meteorol., v. 3, no. 1.

SEIWELL, H. R., 1937. The minimum oxygen concentration in the western basin of the North Atlantic. Papers Phys. Oceanogr. Meteorol., v. 5, no. 3.

SHTOKMAN, V. B., 1937. Applicability of the dynamic method to the analysis of hydrological data in the study of currents in the Caspian Sea. Zh. Geofiz., no. 4.

SHTOKMAN, V. B., 1949. Study of the effect of wind and bottom relief on the distribution of mass and full currents in an inhomogeneous ocean (sea). Tr. Inst. Okeanol. Akad. Nauk SSSR, v. 3.

SHTOKMAN, V. B., 1950. Determination of current velocities and of density distribution in a transverse section of an infinite channel as a function of the wind effect and lateral friction. Dokl. Akad. Nauk SSSR, v. 71, no. 1.

SHTOKMAN, V. B., 1951. Determination of stationary currents and of the field of mass resulting from wind in a baroclinic sea. Tr. Inst. Okeanol. Akad. Nauk SSSR, v. 6.

SHTOKMAN, V. B., 1953a. Some problems concerning the dynamics of sea currents. Izv. Akad. Nauk SSSR, Ser. Geofiz., no. 1.

SHTOKMAN, V. B., 1953b. Simulation of full currents induced by wind in the sea. Izv. Akad. Nauk SSSR, Ser. Geofiz., no. 4.

SHTOKMAN, V. B. and TSIKUNOV, V. A., 1954. Development of full currents in the sea under the effect of wind. Tr. Inst. Okeanol. Akad. Nauk SSSR, v. 9.

SHULEIKIN, V. V., 1945. Convective sea currents in the monsoon field. Dokl. Akad. Nauk SSSR, v. 46, no. 5.

SHULEIKIN, V. V., 1953. Physics of the sea. Izd-vo Akad. Nauk SSSR.

SOMOV, M. M., 1937. Computation of current velocities between hydrological stations of various depths by the dynamic method. Meteorol. i Gidrol., no. 8.

SOULE, F. M., 1939. Consideration of the depth of the motionless surface near the Grand Banks of Newfoundland. Woods Hole Oceanogr. Inst. Coll. Rept., Contrib. 230.

STOMMEL, H., 1950. An example of thermal convection. Trans. Am. Geophys. Union, v. 31, no. 4.

SVERDRUP, H. U., 1938. On the explanation of the oxygen minima and maxima in the ocean. J. Conseil, Conseil Perm. Intern. Exploration Mer., v. 13., no. 2.

SVERDRUP, H. U., JOHNSON, M. W., and FLEMING, R. H., 1942. The oceans, their physics, chemistry and general biology. New York.

SWALLOW, J. C. and WORTHINGTON, L. V., 1959. The deep countercurrent of the Gulf Stream off South Carolina. International Oceanographic Congress, Washington.

TAKANO, K., 1955. An example of thermal convective current. Records Oceanog. Works Japan, v. 2, no. 1.

THOMPSON, E. F., 1939. A rapid method for the determination of dynamic heights (or depths) at successive lowerings at an anchor station. J. Marine Res., v. 2, no. 2.

WATTENBERG, H., 1929. Die Durchlüftung des Atlantischen Ozeans. J. Conseil, Conseil Perm. Intern. Exploration Mer., v. 4.

WÜST, G., 1924. Florida und Antillenstrom. Verhandl. Inst. Meereskunde, New Ser., A, no. 12.

WÜST, G., 1935. Schichtung und Zirkulation des Atlantischen Ozeans. Die Stratosphäre. Wiss Ergeb. Deut. Atlant. Exped. "Meteor" 1925-1927, v. 6, Part 2.

WÜST, G., 1955. Stromgeschwindigkeiten im tiefen- und boden-Wasser auf Grund dynamischer Meteor-Profile der Deutschen Atlantischen Expedition 1925-27. Deep-Sea Res., Suppl. to v. 3.

ZUBOV, N. N., 1929. Computation of the elements of sea currents from hydrological profile data. Zap. po Gidrografii, v. 58.

ZUBOV, N. N. and MAMAEV, O. I., 1956. Dynamic method of computing the elements of sea currents. GIMIZ.
ZUBOV, N. N., 1957. Oceanological tables.

INDEX

Agulhas current, 119
Atmospheric pressure, 5, 80, 83, 86, 90, 101, 102
Atmospheric pressure gradient, 80, 86, 90, 101, 102
Atmospheric stratification index, 83, 84, 86, 89

Baroclinicity, depth of, 36, 37
Baroclinic layer, 30, 73, 76, 77, 93, 98, 100, 101, 109, 111, 145
Baroclinic mass, 155
Baroclinic sea, 9, 12, 30-47
Barotropic fluid, 30
Black Sea, 145

Circulation, convective, 25
Climate, 24-30, 45
Closed system equation, 14
Coefficient of turbulent diffusion, 130-132
Coefficient, turbulent exchange, 15, 16
Continuity equation, 133-139
Convective gradient, 17, 26
Convective circulation, 25
Convective component, 42
Convective current, 29
Coriolis force, 2, 8, 117
Coriolis parameter, 5, 33, 34, 76, 85, 170
Countercurrent, 43, 119, 120
Current, absolute velocity, 17
Current, Agulhas, 119
Current, convective, 29
Current, deep sea, 75-101
Current, geostrophic, 20-23, 38, 47, 169
Current, gradient, 3, 9, 10, 13, 20, 22, 38, 39, 41, 43, 45, 75, 78, 113, 116, 120, 129, 133, 135, 138, 141, 153, 156, 161, 169-173, 175, 182, 186, 190, 192-194, 199-201
Current, Labrador, 43
Current, Oyashio, 43, 176, 183, 186

Current, stratified, 77
Current, surface, 75, 82, 95, 98, 101, 102, 109, 111, 114
Current, tidal, 11, 98
Current velocity, gradient component of, 42, 43, 45, 110-112, 115, 116, 139, 153, 162-164, 169-173, 181, 183, 188, 190, 192, 201
Current, wind driven, 13, 17-23, 27, 28, 30, 34, 35, 73-75, 78, 82, 106, 110-112, 145, 153

Deep-sea current, 75-101
Density field, 9, 22, 26, 73, 192
Density gradient, horizontal, 142
Density gradient, vertical, 32, 33, 37, 105, 110, 111, 116, 117, 125-129
Density model, 73-116
Density, potential, 52, 54
Density, sea-water, 15, 16, 20, 21, 23, 34, 39, 44, 45, 54, 68, 89, 102, 112, 124, 134, 156, 163, 164
Detritus, 121
Discontinuity, 42, 43
Drift component, 17-20
Dynamic computation, 70, 71
Dynamic depth, 147, 148, 182-184, 187
Dynamic height, 55-63, 69, 72, 151, 153, 155, 157-159, 161, 162, 164, 197, 198
Dynamic relief, 45, 58, 69

Evaporation, 24, 25, 26
Electromagnetic current meter, 74
Equation, closed system, 14
Equipotential surface, 30, 170

Fluid, barotropic, 30
Fluid motion equation, 4
Fluid motion, geostrophic, 7, 8, 10

INDEX

Friction, 18, 23, 24, 26-28, 34, 38, 44, 75, 85, 153, 162

Geostrophic current, 20-23, 38, 47, 169
Geostrophic fluid motion, 7, 8, 10
Geostrophic wind, 80, 81, 83, 84, 86
Gradient, atmospheric pressure, 80, 86, 101, 102
Gradient component of current velocity, 42, 43, 45, 110-112, 115, 116, 139, 153, 162-164, 169-173, 181, 183, 188, 190, 192, 201
Gradient, convective, 17, 26
Gradient current, 3, 9, 10, 13, 20, 22, 38, 39, 41, 43, 45, 75, 78, 113, 116, 120, 129, 133, 135, 138, 141, 153, 156, 161, 169-173, 175, 182, 186, 190, 192-194, 199-201
Gradient, horizontal density, 142
Gradient, horizontal pressure, 2, 4, 6, 8-12, 22, 25, 45, 46, 110, 113, 138, 147, 150, 151, 153, 154, 163
Gradient, vertical density, 32, 33, 37, 105, 110, 111, 116, 117, 125-129
Gravity acceleration, 163
Gulf Stream, 43, 119, 120, 127

Homogeneous sea, 9, 34, 147
Horizontal density gradient, 142
Horizontal inhomogeneity, 6, 8, 12, 24
Horizontal pressure gradient, 2, 4, 6, 8-12, 22, 25, 110, 147, 150, 151, 153, 154
Horizontal turbulent friction, 18

Infinitely wide channel, 13-24, 37
Inhomogeneity, 6, 8, 12, 13, 17, 24, 26, 39, 42, 73, 75, 116
Isobaric surface, 1, 6, 8, 9-12, 38, 39, 65, 66, 68, 69, 75, 76, 91, 96, 114, 115, 119, 127, 139-149, 155, 158, 159, 161, 162, 164, 166, 170, 182, 183, 187, 195, 199-201
Isobath, 170

Isohaline, 28, 126, 127, 132, 170, 179
Isopycnal layer, 127-129, 151, 179, 181, 201
Isopycnal motion, 124, 126
Isopycnal surface, 4, 30-32, 38, 40, 42, 121, 124, 126-128, 180
Isostere, 150-154, 167
Isosteric surface, 142, 149
Isotach, 156, 168, 170

Kurile-Kamchatka basin, 113, 129, 174-201
Kuroshio current, 43, 176

Labrador current, 43
Lapse rate, 83
Law of parallelism, 169-173

Measurements, temperature, 48, 49, 54

Oxygen, 118-124, 128, 178
Oyashio current, 43, 176, 183, 186

Parallelism, law of, 169-173
Potential density, 52, 54
Pressure, atmospheric, 5, 80, 83, 86, 90, 101, 102
Pressure gradient, atmospheric, 80, 86, 90, 101, 102
Pressure gradient, horizontal, 2, 4, 6, 8-12, 22, 25, 110, 147, 150, 151, 153, 154
Pycnomere, 124, 125, 127-129, 179

Reference surface, 47, 56, 60, 61, 65, 68, 69, 71, 92, 113, 115-117, 121, 124, 149, 150, 155, 156, 160, 172

Salinity, 27, 28, 30, 39, 44, 48, 49-52, 54, 126, 129-134, 136, 138, 145, 170, 176-179, 186, 187, 190
Sea water density, 15, 16, 20, 21 23, 34, 39, 44, 45, 54, 68, 89, 102, 112, 124, 134, 156, 163, 164

Solenoide, 12, 44, 47, 169-171, 177
Specific volume, 6, 55, 56, 149-152, 158, 161, 164, 166, 167 172, 176, 177, 181
Stratification function, 93, 102, 104-106, 108, 116, 192-194, 201
Stratification index, atmospheric, 83, 84, 86, 89
Stratification, water, 20, 32, 41, 77, 83, 103, 110, 116, 117, 147, 148, 201
Stratified current, 77
Surface current, 75, 82, 95, 98, 101, 102, 109, 111

Tangential wind stress, 15, 30, 33, 34, 36, 74, 78, 80-82, 84, 145
Temperature measurements, 48, 49, 54
Thermal energy, 27
Tidal current, 11, 98
Titration, 50, 51
Transpiration, 24-26
Transverse circulation, 26
Trapezoidal method, 188
Trapezoid formula, 55
Triangulation, 158, 159
Trihedral prisms, 188

Turbulent diffusion, coefficient of, 130-132
Turbulent exchange, 16, 78, 80, 84, 119, 129
Turbulent exchange coefficient, 16
Turbulent friction, 18, 34, 38, 85
Turbulent friction, horizontal, 18

Vertical density gradient, 32, 33, 37, 105, 110, 111, 116, 117, 125-129
Vertical velocity, 73-117
Vityaz', 174

Wind driven current, 13, 17-23, 27, 28, 30, 34, 35, 73-75, 78, 82, 106, 110-112, 145, 153
Wind, geostrophic, 80, 81, 83, 84, 86
Wind stress, tangential, 15, 30, 33, 34, 36, 74, 78, 80-82, 84, 145

'Zero'-surface, 10-12, 26, 31, 35, 45, 46, 70, 75-77, 91, 94, 96, 113-116, 149, 154, 174-195, 199-201
'Zero'-surface, non horizontal, 155-169